D0628923

# Recent Results in Cancer Research 123

# Recent Results in Cancer Research

Volume 113
U. Eppenberger, A. Goldhirsch (Eds.):
Endocrine Therapy and Growth Regulation of Breast Cancer
1989. 26 figures, 17 tables. IX, 92. ISBN 3-540-50456-7

Volume 114
P. Boyle, C.S. Muir, E. Grundmann (Eds.): Cancer Mapping
1989. 97 figures, 64 tables. IX, 277. ISBN 3-540-50490-7

Volume 115
H.-J. Senn, A. Goldhirsch, R.D. Gelber, B. Osterwalder (Eds.):
Adjuvant Therapy of Primary Breast Cancer
1989. 64 figures, 94 tables. XVI, 296. ISBN 3-540-18810-X

Volume 116
K.W. Brunner, H. Fleisch, H.-J. Senn (Eds.):
Bisphosphonates and Tumor Osteolysis
1989. 22 figures, 6 tables. IX, 78. ISBN 3-540-50560-1

Volume 117
V. Diehl, M. Pfreundschuh, M. Loeffler (Eds.):
New Aspects in the Diagnosis and Treatment of
Hodgkin's Disease
1989. 71 figures, 91 tables. XIV, 283. ISBN 3-540-51124-5

Volume 118
L. Beck, E. Grundmann, R. Ackermann, H.-D. Röher (Eds.):
Hormone-Related Malignant Tumors
1990. 83 figures, 91 tables. XI, 269. ISBN 3-540-51258-6

Volume 119
S. Brünner, B. Langfeldt (Eds.):
Advances in Breast Cancer Detection
1990. 53 figures, 51 tables. XI, 195. ISBN 3-540-52089-9

Volume 120
P. Band (Ed.): Occupational Cancer Epidemiology
1990. 23 figures, 59 tables. IX, 193. ISBN 3-540-52090-2

Volume 121
H.-J. Senn, A. Glaus (Eds.):
Supportive Care in Cancer Patients II
1991. 49 figures, 94 tables. XIV, 465. ISBN 3-540-52346-4

Volume 122
W.J. Eylenbosch, M. Kirsch-Volders, A. Deleener,
J. Weyler (Eds.): Primary Prevention and Cancer
1991, 16 figures, 28 tables. IX, 130. ISBN 3-540-53704-X

J.W. Oosterhuis H. Walt
I. Damjanov (Eds.)

# *Pathobiology of Human Germ Cell Neoplasia*

With 48 Figures and 10 Tables

Springer-Verlag
Berlin Heidelberg New York
London Paris Tokyo
Hong Kong Barcelona
Budapest

Prof. Dr. J. Wolter Oosterhuis
Dr. Daniel den Hoed Cancer Center
Groene Hilledijk 301
NL-3075 EA Rotterdam, The Netherlands

Dr. Heinrich Walt
Department of Gynecology and Obstetrics
Research Division of Gynecology
University Hospital
CH-8091 Zürich, Switzerland

Prof. Dr. Ivan Damjanov
Department of Pathology
Jefferson Medical College
Thomas Jefferson University
132 South 10th Street
Philadelphia, PA 19107, USA



ISBN 3-540-53928-X Springer-Verlag Berlin Heidelberg New York
ISBN 0-387-53928-X Springer-Verlag New York Berlin Heidelberg



Typesetting: Best-set Typesetter Ltd., Hong Kong
25/3130-543210 – Printed on acid-free paper.

# *Preface*

Germ cell tumors have been fascinating pathologists for decades. It was not until recently, however, that major advances in the chemotherapy of malignant germ cell tumors focused the attention of the entire medical community on these neoplasms. Once lethal for most patients, these tumors have become a prototype of successful chemotherapy. Basic scientists have become interested in the subject since the cloning and characterization of the cell lines of human germ cell tumors was introduced about 10 years ago. Primarily meant as tools to study human developmental biology, these cell lines are also employed to investigate the biology of human germ cell neoplasia.

It is certain that students of human germ cell neoplasia have much to learn from developmental biologists, in particular those who use murine teratocarcinoma as models. The reverse is probably also true. To stimulate this mutual learning process, the symposium Pathobiology of Human Germ Cell Neoplasia was organized in August 1989, in Groningen, as a satellite meeting of the 11th Congress of the International Society of Developmental Biologists held in Utrecht. The present volume is based on this meeting and provides information to the public not directly concerned with research on germ cell tumors about the recent progress made in the field.

The financial support of our sponsors is gratefully acknowledged: J.K. de Cock-Stichting, Groningen; Stichting Pathologisch Anatomisch Laboratorium, Groningen; Koninklijke Smilde B.V., Heerenveen; The Netherlands Cancer Foundation, Amsterdam; Stichting Kinderoncologie Groningen, Groningen; Stichting Werkgroep Interne Oncologie, Groningen; Nederlandse Middenstands Bank, Groningen; and Bristol Myers, Weesp.

We thank the authors for their excellent contributions to this book. The opportunity given to us by Professor E. Grundmann to publish the proceedings of the meeting in the series *Recent Results in Cancer Research* is highly appreciated. Finally we thank Silwia Nijdam for her enthusiastic secretarial help.

We hope that the readers of this volume may experience some of the excitement that was ours at the meeting in Groningen.

J.W. Oosterhuis
H. Walt
I. Damjanov

*Contents*

# *List of Contributors**



---

* The address of the principal author is given on the first page of each contribution.

[1] Page on which contribution begins.

# Overview

## *Pathobiology of Human Germ Cell Neoplasia*

I. Damjanov

Department of Pathology and Cell Biology, Thomas Jefferson University,
Jefferson Medical College, Philadelphia, PA 19107, USA

### Introduction

The term *germ cell neoplasia* is used today histogenetically to define a rather heterogeneous group of tumors whose origin and morphogenesis can be best explained by postulating that they all originate from germ cells. Since most germ cells reside in the gonads, germ cell tumors are predominantly located in the testis and the ovary. However, morphologically identical tumors occur in extragonadal sites as well, raising at least some doubts about an obligatory germ cell derivation of these neoplasms. Is it possible that some of the extragonadal tumors are not of germ cell origin but rather derived from developmentally pluripotent embryonic or even extraembryonic cells displaced to these sites during the development? Are the stem cells in all the germ cell tumors of man, irrespective of their origin, identical or do they bear unique features imprinted during their development in different sites? Are the stem cells of a monomorphic, developmentally nullipotent embryonal carcinoma (EC) equivalent to the stem cells of a teratocarcinoma (mixed germ cell tumor)? Developmentally they are obviously different since the nullipotent EC cannot differentiate into many other cells and tissues, whereas the stem cells of teratocarcinoma give rise to various somatic and extraembryonic cell forms. However, since these cells are morphologically indistinguishable from one another and may express the same cell surface antigens (Andrews 1988), one would be inclined to conclude that they are indeed identical. However, what should take precedence for taxonomic purposes: the developmental potential or the cell phenotype? Could the chromosomal and genetic markers help out in this respect and what would represent the conclusive evidence that would definitively resolve all the dilemmas?

The relationship of EC to seminomas and carcinoma in situ (CIS) could provide some other important clues. Does EC develop directly from intratubular CIS or through an intermediary stage equivalent to seminoma (Damjanov 1989)? Can seminomas differentiate or even progress to EC

Recent Results in Cancer Research, Vol. 123

as hypothesized on the basis of DNA flow cytometry studies (Oosterhuis et al. 1989)? Finally, the vast experience with experimental teratocarcinomas (Stevens 1967; Damjanov and Solter 1974; Silver et al. 1983) is obviously highly relevant for the understanding of the histogenesis and the biology of human germ cell neoplasia and has to be considered as experimental correlates for the phenomena that cannot be directly studied directly in humans.

In this article I will touch upon some of the historic and contemporary concepts important for the understanding of the pathobiology of human germ cell tumors and will briefly review the recent contributions made by studying the human and animal germ cell tumors and cell lines derived from them.

## Histogenesis

As remarked many years ago by Johnson (1953), the histogenesis of tumors is like paternity – a matter of faith. Most of us know who our mother is, but few of us ask for sacrosanct proofs about the identify of our father. Likewise, since none of us was there "when the picture was painted" [a paraphrase of Francis Crick (1988), who in his book of reminiscences quoted John Minton about artistic creation: "The important thing is to be there when the picture is painted"], we are prone to believe the most plausible or the most fashionable explanation of the time in which we live. Today we take for a fact that the so-called germ cell tumors originate from germ cells, but we forget that the final proof is not yet in and that at least some extragonadal germ cell tumors are probably of somatic cell rather than germ cell origin.

### *Testicular Tumors*

Current belief that germ cell tumors originate from the germ cells in the seminiferous tubules could be traced to the classical paper by Ewing (1911), who reviewed the older literature, summarized his own observations, and argued most convincingly against the histogenetic explanations that were fashionable in those days. As if Ewing were into a game of name-dropping, his paper reads like an eponymic dictionary attesting to the fact that the germ cells have preoccupied the minds of some of the greatest pathologists of the golden era of anatomic pathology. Virchow, Paget, Waldayer, Wilms, Chevassu, Curling, Wilms, Walthard, Malassez, McCallum, Warthin, Langhans, Pick, and von Hansemann are among those whose reports Ewing reviewed and discussed, not to mention some other pathologists, who might have been as important those days, but whose names have not been eponymically linked to a specific disease and are, thus, less recognizable.

Ewing (1911) reviewed seven theories proposed to explain the origin of teratoma of the testis and its derivatives, i.e., tumors included today under the name of germ cell neoplasia.

*Theory of Metaplasia*

This theory, favored by Virchow and perpetuated in many textbooks of those days, is of little interest today. It proposes that the epithelial cells of teratomas originate from the germinal epithelium and the stroma from the testicular connective tissue cells. Obviously the monoclonality of human tumors and the malignant transformation of single cells have not been concepts familiar to the father of cellular pathology.

*Theory of Fetal Inclusions*

This theory could be traced to the first recorded case of testicular teratoma reported in 1696 by Saint Donat (Ewing 1911). According to this theory, teratomas should be considered "foetal monstruosities" rather than tumors. Although some of the retroperitoneal and sacrococcygeal teratomas of the newborn contain fetal parts (Nadimpalli et al. 1989), suggesting that they are indeed "fetus in feto," the fetal inclusion theory is a very unlikely explanation for testicular teratoids, which Ewing (1911) interpreted as "the offspring and not the brothers of the host."

*Theory of Partial Hermaphroditism*

This theory was promoted in several forms, which all implicate intratesticular or paratesticular oocytes as the progenitors of teratomas. Waldayer (cited by Ewing 1911) starts his hypothesis with the premise that the ovary and testis begin from a sexually indifferent stage primordium. Hence, he proposes that the female germ cell development may occur in the testis because of the incomplete suppression of the female characteristics during the abnormal development. From these "primordial" ova, which Waldayer saw as large clear cells with large nuclei and which he believed to remain dormant in some adult testes, one could obtain tumors through parthenogenesis. Ewing (1911) does not completely dismiss the parthenogenetic theory of Waldayer and states that since "teratoma are not perfect embryos, it is conceivable that an incomplete form of parthenogenesis may account for the rudimentary growth of embryomata." Although there is currently no proof that there are any oocytes in the testis of normal males, activation of testicular germ cells to form embryoids and somatic structures could represent an equivalent of parthenogenetic activation of male germ cells. Experimentally, the parthenogenetic activation of murine oocytes could be achieved by various chemicals and even viruses (Kaufman, 1983), and since some of these stimuli may also be carcinogenic, testicular equivalents of ovarian parthenogenesis deserve further study.

The partial hermaphroditism theory championed by Langhans (cited by Ewing 1911) needs no serious consideration and I am retelling it here only to illustrate how some histogenetic explanations defy the limits of imagination. According to this theory, the tumors originate from ovarian rudiments that are occasionally included in the testis. Oocytes developing in these rudiments are fertilized by the spermatozoa and give rise to embryos which become disorganized and form tumors with embryoid features. This hypothesis, obviously, does not have too many adherents, although it is based on an important observation – that most testicular germ cell tumors occur following the onset of spermatogenesis at puberty. It is, however, unlikely that the postpubertal occurrence of tumors has anything to do with the maturation of spermatozoa and their acquisition of a fertilizing potential.

### *Theory of Fertilization of Polar Body*

This theory, based on misinterpreted data from invertebrate embryology, could be considered in the pathogenesis of ovarian teratomas. However, it is hard to imagine that it could be relevant for testicular germ cell tumors if for no other reason than the fact that nucleated polar bodies do not form during spermatogenesis.

### *Theory of an Isolated Blastomere*

This theory proposed by Marchand and Bonnet implies that the teratoid tumors originate from the blastomeres segregated during early embryonic development. Ewing (1911) dismissed this hypothesis as improbable. This theory, representing the modified embryonic rest theory of Conheim, subsequently had numerous followers and served as the basis for the so-called British histogenetic theory of teratomas (Collins and Pugh 1964). Since experimental teratocarcinomas can be produced from early embryonic cells transplanted to extrauterine sites (Stevens 1970a; Solter et al. 1970), it obviously has some merit. However, like other partially acceptable theories, it probably does not account for the histogenesis of all germ cell tumors. Moreover, the credibility of this theory could be increased by combining it with germ cell theory. Teratomas could, thus, form from germ cells that give rise to embryonic cells, which in turn are the direct progenitors of tumor cells. Once the blastomeres are formed, it is relatively easy to imagine that these developmentally pluripotent cells could differentiate into fetal or mature somatic tissues or cells corresponding to extraembryonic appendages and membranes, i.e., cells equivalent to yolk sac or trophoblast.

Expounding on the blastomeric theory, Wilms (1896, 1898) proposed a very important concept that, in a slightly modified form, has been applied to several tumor systems. Wilms proposed that the structure and histologic appearances of teratomas depend on the stage of embryonic differentiation at which the originating cell complex is isolated. Translated into today's terminology, Wilms' concept implies that the developmental potential of the embryonic cells, i.e., the stage of development at which they have undergone malignant transformation, determines the histologic features of the evolving tumor. Experimental studies on mouse teratocarcinomas have given strong support to this concept. Thus, it has been shown that the stem cells of mouse teratocarcinomas typically correspond to embryonic cells forming the inner cell mass of the blastocyst (Solter and Damjanov 1979). Accordingly, these cells can differentiate into all the cell lineages derived from the inner cell mass. These mouse EC cells, like their normal equivalents in the inner cell mass of the blastocyst, have segregated themselves from trophectoderm and cannot give rise to trophoblast lineages stemming from the trophoblast (Gardner 1983). In contrast, the majority of human EC cells can still give rise in vivo to trophoblastic cells (Ulbright and Loehrer 1988) and do, by inference, correspond to embryonic cells that are less differentiated than the mouse EC or the inner cell mass of the blastocyst.

### *Theory of Wolffian or Müllerian Duct Origin*

This theory has received limited attention since there is no evidence that complex structures or developmentally pluripotent cells could originate from remnants of these embryonic structures. It is interesting to note that Cavazanni, who according to Ewing (1911) was the major proponent of this theory, considered the testicular tumors of early life as usually benign and of different histogenesis than those that originate in later life. Again, although the theory has no contemporary appeal, it is well accepted that most testicular tumors of early life are benign (Mahour et al. 1974; Kaplan et al. 1988). Also, it is quite possible, if one were to judge from the cytogenetic studies (Oosterhuis et al. 1989), that childhood yolk sac tumors may be of different origin or histogenetically distinct from tumors in older persons. The fact that the testicular tumors of infancy and childhood are not associated with CIS in the peritumoral seminiferous tubules (Manivel et al. 1988, 1989) also indicates that these tumors might have a different histogenesis than the tumors of adult testis.

### *Adrenal Rest Theory*

This theory has been based on the observation that adrenal rests may occur in the testis. Although adrenal rests may be found in the testes and may give rise to tumors (Rutgers et al. 1988), it seems unlikely that the teratoid tumors would originate from adrenal cells. In this context it is of interest to note that Upadhyay and Zamboni (1982) and Zamboni and Upadhyay (1983) have noted aberrant germ cells frequently in murine fetal adrenals. Thus, even if adrenal rests included in the testis were to give rise to teratomas, a hypothetical and so far undocumented event, the tumors would have more likely been derived from ectopic germ cells included in these embryonic rests than the adrenal cortex itself.

In the final analysis of Ewing's paper, one cannot be but amazed at how exhaustive and all encompassing his discussion of the histogenesis of germ cell tumors was. Furthermore, as I tried to show, all these theories, irrespective of their present-day relevance and credibility, seem to be based on valid observations. In most instances one can reconstruct the original thinker's ideation that led to the formulation of the hypothesis and even relate the old concept to some recent scientific discovery.

### *Ovarian Germ Cell Tumors*

Ovarian germ cell tumors and their histogenesis can be conceptually linked to parthenogenesis much easier than the testicular tumors. Parthenogenetic activation of ova is a well-documented phenomenon in mice (Stevens and Varnum 1974). Thus, it is quite probable that such an event could take place in the human ovaries as well. Human ova collected for in vitro fertilization may show a single pronucleus suggestive of parthenogenetic activation (Plachot et al. 1988). Although this occurs in less than 2% of oocytes in vitro, it nevertheless supports the notion that the parthenogenetic activation could occur in human intraovarian oocytes as well. The aspects of oogenesis that predispose to parthenogenesis in mammals are not understood (Kaufman 1983). Without the contribution of male chromosomes, activated ova have, however, a limited developmental potential (Solter 1988).

The occurrence of parthenogenones in strain LT mice is genetically predetermined, but under hormonal control (Eppig et al. 1977). The occurrence of ovarian teratomas after puberty suggests that hormonal factors may be of importance in the pathogenesis of human tumors as well. On the other hand, epidemiologic studies do not strongly support the hypothesis that genetic factors predispose to ovarian germ cell tumorigenesis. Familial cases of ovarian teratoma have been described but are apparently rare (Gustavson and Rune 1988). Bilateral ovarian teratomas occur in about 5% of patients and multiple distinct teratomas even in the same ovary have been described (Carritt et al. 1982). These provide additional evidence that some patients have a higher "predilection" to tumorigenesis but the reasons for it remain obscure. A case of ovarian teratoma in a Li-Fraumeni cancer family in which five members had testicular tumors (Hartley et al. 1989) indicated that the ovarian tumorigenesis may be related to the same factors that cause testicular tumors.

In analogy with the preinvasive stages or CIS of testicular tumors (Giwercman et al. 1988), one could postulate that invasive ovarian germ cell tumors are preceded by carcinoma in situ or that one could find evidence of early parthenogenetic cleavage of oocytes in the ovary adjacent to the germ cell tumors. We are unaware of any systematic studies in which this problem was explored. Binucleated oocytes can occasionally be seen within primordial follicles in ovaries with dysgerminoma, and some follicles resemble the atypical follicles deficient in the number of granulosa cells described by Eppig (1978) in the ovaries of LT mice. However, in contrast to the reports on LT mice, there are no reports of intraovarian cleavage stage human embryos.

### *Extragonadal Tumors*

Extragonadal germ cell tumors could originate from aberrant germ cells or from inclusions of the twin fetus (see discussion in Stephens et al. 1989). Sacrococcygeal teratomas in the newborn often have fetiform features, and of all the teratoid tumors these are the most likely to be derived from blastomeres or embryonic and fetal cell inclusions. There is, however, also a possibility that some extragonadal teratomas, especially those in the anterior mediastinum, could have originated from the inclusions of the fetal yolk sac, a temporary structure that usually involutes before birth. However, since the yolk sac contains primordial germ cells and since the displaced yolk sac can give rise to teratoid tumors and yolk sac carcinoma in rodents (reviewed by Vandeputte and Sobis 1988), one cannot exclude the possibility that some of the extragonadal teratoid tumors are of yolk sac origin. In this context it is notable that the yolk sac carcinoma represents the most prominent malignant component of extragonadal germ cell tumors (Billmire and Grossfeld 1986). Cytogenetic studies are also consistent with the somatic cell derivation or premeiotic origin of these tumors (Mann et al. 1983). The fact that the tumors are often near-diploid, rather than near-tetraploid or near-triploid-like testicular germ cell tumors (Dal Cin et al. 1989), also points to a different histogenesis. On the other hand, the presence of the isochromosome 12p points to some common features shared by germ cell tumors from different anatomic sites.

Anterior mediastinal germ cell tumors occur at an increased rate in patients with Klinefelter's syndrome (reviewed by Lachman et al. 1986). This paradoxical association of germ cell tumors with a chromosomal syndrome characterized by underdeveloped testes and a lack of germ cells in the seminiferous tubules defies explanation. It is possible that the elevated levels of gonadotropins in this syndrome play a role in tumorigenesis. It is also conceivable that the primordial germ cells in these patients are more prone to migrate to extragonadal sites or alternatively that they survive more likely in those sites than in the testes. Sex chromosome abnormalities may be one of the predisposing factors facilitating the formation of germ cell tumors in extragonadal sites (Dexeus et al. 1988).

The tumors originating in extragonadal sites have the same morphology as those in the gonads. A recently established stem cell line, called NCC-IT, derived from a mediastinal teratocarcinoma of a 24-year-old male (Teshima et al. 1988), could serve for further studies of these tumors. This cell line is developmentally pluripotent and is capable of embryonic differentiation. Comparison of cell lines described from gonadal and extragonadal germ cell tumors could provide some clues in the future about the histogenesis and basic nature of these tumors.

### *Tumors of Dysgenetic Gonads*

Germ cell tumors are common in dysgenetic gonads (Robboy et al. 1982). The majority of such tumors were reported in individuals carrying a Y chromosome (Kingsbury et al. 1987; Robboy et al. 1982). Thus, paradoxically, although the abnormal gonads usually contain fewer germ cells than the normal gonad, the germ cells that survive in such an inimical environment tend more often to undergo malignant transformation than the cells located in their normal habitat. One could argue that the cells surviving in adverse conditions have been preselected and, thus, have a growth advantage or alternatively that they survive because they are already "transformed."

Dysgenetic gonads may contain gonadoblastomas (Damjanov et al. 1975), a preinvasive form of germ cell neoplasia. Such gonadoblastomas are probably present from the earliest stages of gonadogenesis and may be histologically identified even in infancy (Damjanov and Klauber 1980). Invasive tumors usually arise from such gonadoblastomas only after a prolonged latent period, but the factors promoting the development of malignancy are not understood.

### *Developmental Stage of Germ Cells That Give Rise to Tumors*

Extensive chromosomal studies on human germ cell tumors (reviewed by Mutter 1987) indicate that most benign teratomas originate from postmeiotic oocytes. However, several tumors showed heterozygous centromeric heteromorphism, indicating that some might have arisen from oocytes that have not completed the meiosis 1. In one patient with seven teratomas, chromosomal studies disclosed that the tumors differed from one another and that they could have originated from oocytes parthenogenetically activated at different stages of meiotic maturation (Carritt et al. 1982). Immature teratomas of the ovary are frequently aneuploid, some cases originating from premeiotic and some from postmeiotic oocytes (Ohama et al. 1985). Thus, one could conclude that the tumors originate from either premeiotic or postmeiotic oocytes (Parrington et al. 1984). Tumors derived from postmeiotic oocytes tend to be more often benign, whereas those originating from premeiotic oocytes are often malignant. This fact, coupled with the fact that the immature and malignant solid teratomas of the ovary occur in younger individuals than cystic teratomas (Malogolowkin et al. 1989), suggests a different pathogenesis and even origin of prepubertal and postpubertal tumors. It is possible that the malignant teratomas are initiated during fetal gonadal development and derived from primordial germ cells. On the other hand, the dermoid cysts (benign cystic teratomas) are mostly tumors of oocytes that have reached greater meiotic maturity and are stimulated by hormonal factors to undergo further development.

Ovarian teratomas in LT mice originate from oocytes that have completed meiosis 1 but not meiosis 2 (Eppig et al. 1977). These murine tumors are thus a better model for human cystic teratomas than solid immature teratomas of young girls. Like the human cystic teratomas, the LT mouse tumors are almost all benign and malignant teratomas (teratocarcinomas) occur only exceptionally (Stevens and Varnum 1974).

The origin of mouse testicular germ cell tumors can be traced to the earliest stages of fetal gonad formation (Stevens 1967). Mouse tumors have an XY karyotype and are, thus, premeiotic. In analogy with murine models, one could postulate that human tumors have their origin in prenatal life as well. However, such studies on human material can be performed only retrospectively. The occurrence of CIS in the prepubertal boys with abnormal gonadal development (Muller et al. 1984; Cortes et al. 1989) is the strongest indication that the human germ cell tumors could ultimately be traced to intrauterine stages of gonadal development. Even so, it remains to be determined whether the cells of origin correspond to primordial germ cells or to more differentiated spermatogenic cells.

Chromosomal studies have disclosed an aneuploidy of most testicular tumors (Castedo et al. 1989a,b). This does not provide any clues about the developmental stage of the germ cells from which the tumor originated except that such cells are most likely premeiotic. Marker studies with monoclonal antibodies to the stage-specific embryonic antigen 3 (SSEA-3), a marker of human embryonal carcinoma cells (Shevinsky et al. 1982; Damjanov et al. 1982) and placental isozyme of the alkaline phosphatase (Paiva et al. 1983; Manivel et al. 1987), have shown that the fetal germ cells express these antigens like the embryonal carcinoma cells. This would suggest that the malignant stem cells of germ cell tumors have retained some features of embryonic germ cells. It is arguable whether this should be taken as evidence that the tumors indeed originated from the fetal or primordial germ cells. It also remains to be determined whether primordial germ cells could remain sequestered, not undergoing maturation during development in the testicular seminiferous tubules. One could hypothesize that such primoridal germ cells would be less prone to respond to the normal regulatory influences of Sertoli's cells and the various hormonal and endocrine and paracrine stimuli regulating the normal spermatogenic cells in the seminiferous tubules.

Extragonadal germ cell tumors could theoretically originate from aberrant germ cells that have lodged in ectopic sites while migrating from the yolk sac to the gonadal anlage during fetal development. Ectopic germ cells have been demonstrated histologically in the mouse (Francavilla and Zamboni 1985) and human fetuses (Falin 1969). Zamboni and Upadhyay (1983) have shown that the ectopic germ cells, irrespective of the genetic sex of the fetus, develop like the normal oocytes in the ovary. However, most ectopic germ cells die in the late stages of pregnancy or early postnatal life. Theoretically, it is nevertheless possible that some germ cells will survive and, thus, could give rise to germ cell tumors. It is not known whether the germ cells with an abnormal chromosome complement, like those in Klinefelter's syndrome, survive longer in ectopic sites than normal karyotypic germ cells. It is also notable that no ectopic germ cells have been demonstrated in human fetuses outside the abdominal cavity (Falin 1969). This, coupled with chromosomal data (see Dal Cin et al. 1989), casts doubt on germ cell derivation of mediastinal tumors, although they appear morphologically indistinguishable from their gonadal equivalents.

## Classification of Germ Cell Tumors

The classification of testicular, and to an extent ovarian, germ cell tumors has been in a state of flux for many years, reflecting the uncertainties about the histogenesis of these tumors. Major controversy still exists about the exact nosology of various morphologic entities, the interrelatedness of the tumors, the diagnostic criteria, and the relevance of the morphologic data for the clinical treatment.

### *Histogenetic Classification*

Essentially all the histogenetic classifications of testicular germ cell tumors derive from the concepts developed by Friedman and Moore (1946) and further refined and disseminated by Dixon and Moore (1952). According to the basic tenets of this theory, these tumors originate from germ cells which may develop along two unrelated pathways. Thus, seminomas (dysgerminomas) recapitulate some aspects of spermatogenesis and are distinct from the other tumors. Nonseminomatous germ cell tumors (NSGCTs) represent descendants of pluripotent embryonic cells that were formed from the germ cell through parthenogenetic activation. This concept (Fig. 1) is strongly supported by experimental data obtained in animals (reviewed by Pierce and Abell 1970).

The major problems with this classification lie with the nosology of seminomas. The concept that seminomas represent spermatogenic tumors is not compatible with the fact that morphologically similar tumors,

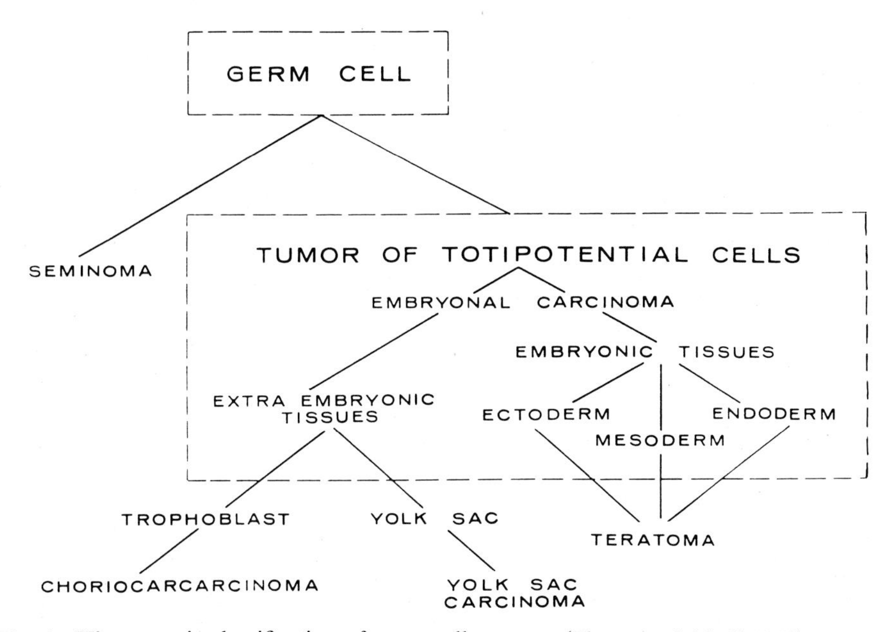


**Fig. 1.** Histogenetic classification of germ cell tumors. (Pierce and Abell 1970)

dysgerminomas, occur in the ovary, an organ that has no spermatogenic potential. Testicular seminomas and ovarian dysgerminomas are morphologically indistinguishable. Thus, it is reasonable to assume that these tumors originate from the same percursors and, even more importantly, that they are composed of the same cell types. Extragonadal seminomas/dysgerminomas belong to the same morphologic tumor type although there is no evidence that spermatogenesis could occur outside of the testis.

Skakkebaek et al. (1987) have argued quite convincingly that seminomas should not be considered spermatogenic tumors, and have histogenetically included them together with other NSGCTs. All these tumors are derived from intratubular germ cells that pass through a preinvasive CIS stage. Numerous mixed germ cell tumors that contain both seminoma and NSGCTs components also favor a close relatedness of seminomas and other germ cell tumors. Finally, the DNA fluorocytometry data (Oosterhuis et al. 1989), discussed in greater detail elsewhere in this monograph, are also quite consistent with the concept that links seminomas and NSGCTs through a common histogenetic pathway (Fig. 2). Spermatocytic seminoma is distinct from classical seminoma, and is the only testicular tumor that shows some similarity with spermatogenic cells (Skakkebaek et al. 1987).

There is a general consensus that all testicular tumors originate from intratubular germ cells (Fig. 2). It has not been resolved whether the tumors derive clonally from a single cell that has undergone malignant transformation or whether the process of tumorigenesis is multifocal. Testicular tumors of 129 strain mice may be multiple and the process seems to be multifocal

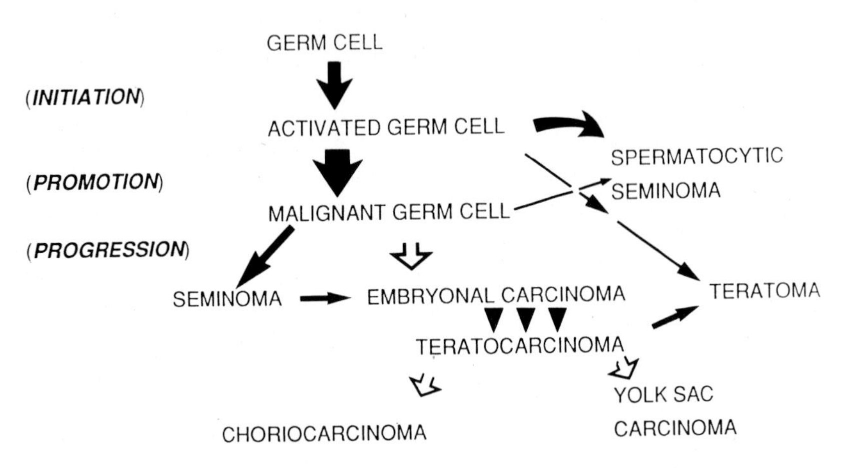


**Fig. 2.** Revised histogenetic classification of testicular germ cell tumors

(Stevens 1967). The genetic and/or environmental factors apparently activate more than one primordial germ cell to undergo tumorigenesis (Stevens 1970b).

The development of murine testicular tumors can be traced to the earliest stages of seminiferous tubule formation in the fetal gonad shortly after the primordial germ cells have entered into the genital ridge (Stevens 1967). The mouse tumors are, thus, fetal in origin and are derived from primordial germ cells.

We have shown that the primoridal germ cells in the 12- to 14-gestational-day-old fetal gonad express SSEA-1, the well-known cell surface epitope of mouse embryonal carcinoma cells (Fox et al. 1981). It is of interest to note that SSEA-1, disappears from the fetal germ cells after day 14 of pregnancy, coincidentally with the loss of the tumorigenic potential of these cells transplanted to the testes of adult hosts (Stevens 1970b). It would, thus, appear that there is direct continuity between the SSEA-1-positive primordial germ cells and the stem cells of teratocarcinomas.

In human fetuses the fetal germ cells express SSEA-3, another carbohydrate marker that is typically found on human embryonal carcinoma cells (Damjanov et al. 1982). This finding could be interpreted as evidence of retrogression of tumor cells to a fetal phenotype or persistence of certain fetal germ cell features on the tumor stem cells. The fetal, i.e., prenatal, initiation of human testicular germ cell tumors is a distinct possibility that deserves to be explored. The finding of CIS in prepubertal males (Müller et al. 1984), i.e., many years before the peak incidence of testicular germ cell tumors, also indicates that the initial tumorigenic event could have occurred in infancy and even prenatally.

The transition from CIS to invasive carcinoma is marked by a long latency (reviewed by Giwercman et al. 1988). It is not known whether the tumor cells have the potential to grow invasively from the beginning or whether the acquisition of invasiveness is the result of a "second hit," which would correspond to the cocarcinogenic effects of promoters in other tumor systems of multistep carcinogenesis (Pitot et al. 1988). The ploidy studies of testicular tumors (Oosterhuis et al. 1989) are consistent with at least a two-step event leading to the transition of initially polyploidized noninvasive CIS cells into invasive cancer.

In the scheme of testicular germ cell tumors presented in Fig. 2, we postulate several discrete steps which could be designated as *initiation*, equivalent to parthenogenetic activation of germ cell; *promotion*, which results in recognizable intratubular carcinoma cells with a tetraploid chromosome number, but no obvious invasiveness; and *progression* of CIS into an invasive neoplasm. Theoretically, CIS could give rise to either seminoma or embryonal carcinoma. Embryonal carcinoma cells are the malignant equivalents of pluripotent embryonic cells, capable of forming fetal and adult somatic cells and tissues as well as giving rise to extraembryonic cells of the placenta and the yolk sac. In this scheme seminomas represent either monophasic tumors composed of terminally differentiated cells or an intermediate developmental stage of embryonal carcinoma. The latter possibility is favored by Skakkebaek et al. (1987) and supported by ultrastructural studies (Srigley et al. 1988) and DNA ploidy data (Oosterhuis et al. 1989). It remains to be determined whether seminoma is a prerequisite stage for formation of other NSGCTs or whether embryonal carcinoma can form directly from CIS.

Numerous mixed germ cell tumors composed of seminoma and embryonal carcinoma support the contention that seminoma is closely related to embryonal carcinoma. Findings of trophoblastic giant cells in at least 15% of otherwise monomorphic seminomas (Mostofi et al. 1987) indicate that not all seminomas are composed of terminally differentiated cells. The fact that there are several well-characterized embryonal carcinoma cell lines propagated in vitro, and that nobody was able to grow seminomas in culture so far, is compatible with the suggestion that the embryonal carcinomas have progressed further than seminomas and are, therefore, more malignant.

Embryonal carcinoma cells may be developmentally nullipotent and thus form monomorphic tumors. On the other hand, some of these cells may undergo somatic and extrasomatic (extraembryonic) differentiation and thus form teratocarcinomas. As evidenced by the chemotherapy data (Oosterhuis et al. 1983; Fosså et al. 1989), teratocarcinomas may lose their malignancy and become benign teratomas. On the other hand, some teratomas are benign tumors ab initio and there is no evidence that such tumors have passed through a stage equivalent to teratocarcinoma. We postulate that these teratomas are derived from germ cells that have been activated to undergo parthenogenesis but have never been exposed to the "second hit" to become malignant. It is even possible that some teratomas are derived from dystopic extratubular germ cells, because the seminiferous tubules adjacent to these tumors do not contain CIS (Manivel et al. 1989).

Yolk sac carcinomas of infants could also originate from germ cells that have not passed through the polyploidization and CIS formation (Oosterhuis et al. 1989). In infants the peritumoral seminiferous tubules do not contain CIS (Manivel et al. 1988) and the clinical course is indolent, suggesting major differences between childhood and adult testicular yolk sac tumors.

Figure 2 shows unequivocally that spermatocytic seminoma is not related to the classical seminoma and is actually quite distinct from all other germ cell tumors. As suggested by Skakkebaek et al. (1987), spermatocytic seminoma is most likely the only truly spermatocytic tumor, and its cells are equivalent to spermatogonia or spermatids. The exact nosology of spermatocytic seminoma. nevertheless, remains conjectural, especially in view of recent reports of tumors composed of spermatocytic seminoma and malignant somatic cells (True et al. 1988; Floyd et al. 1988). Thus, although the majority of spermatocytic seminomas are benign and could have originated from germ cells that have not undergone malignant transformation, some tumors have a malignant potential. There are no fully acceptable explanations for the occurrence of heterologous somatic cells in spermatocytic seminomas with "sarcomatous transformation" (True et al. 1988; Floyd et al. 1988).

Pure choriocarcinoma and yolk sac carcinoma of adult testis are listed in Fig. 2 as derived from pluripotent embryonal carcinoma cells, i.e., the stem cells of teratocarcinoma. Most of the trophoblastic or yolk sac components in NSGCTs of the testis represent terminally differentiated human chorionic gonadotrophin (HCG) or α-fetoprotein producing nonproliferating cells (Mostofi et al. 1987). Some of these cells have nevertheless retained their malignant potential and form the pure choriocarcinomas of the testis. These tumors are extremely rare (Ulbright and Loehrer 1988) and many probably represent the overgrowth of trophoblastic cells combined with destruction and obliteration of all other parts of the original germ cell tumor. It is questionable whether pure yolk sac carcinomas occur in adult testis at all, thus making a systematic study of these problems very difficult. Yolk sac carcinomas of the ovary and extragonadal sites are malignant (Ulbright et al. 1986) and clinically quite distinct from testicular yolk sac carcinomas. It remains to be determined whether they have the same histogenesis as testicular tumors.

### *Histologic Descriptive Classification*

The classification of testicular tumors based on the listing of components recognized on histologic examination has been promoted by the British Testicular Tumor Panel (Pugh 1976). This approach has been less popular in the United States, but is widely used in Europe. One may object to a somewhat arbitrary terminology, but it is reproducible, relatively simple, and reliable. One may object that there are no biological or clinical justifications for some aspects of this classification. Thus, e.g., there are no reasons for lumping together *teratocarcinoma*, a tumor whose stem cells are equivalent to embryonal carcinoma, and the *teratoma with malignant transformation*, whose malignancy resides in proliferation of malignant somatic cells. Other concerns have been voiced, but these are all minor quibbles in view of the proven practicality of this approach. Since this classification provides the most systematic means for correlating with clinical findings and has the best interobserver correlation, one could predict that it will be used for many years to come.

### *Clinical Classification*

Recent advances in biochemical diagnosis of germ cell tumors and the most successful treatment of these tumors based on surgery, drug, and/or radiation regimens has dramatically changed the outlook of the pathology of testicular germ cell tumors. For practical reasons it is still important for the pathologist to ascertain histologically whether a tumor is benign or malignant. Marker studies have considerably improved the morphologic diagnosis. Nevertheless it is not uncommon that the histologic diagnosis, especially that of benign tumors, may be challenged or modified if the serologic findings provide evidence of elevated serum HCG or α-fetoprotein (Javadpour 1986). For the sake of therapy and follow-up the germ cell tumors of adult testis are, thus, in many major institutions classified on the basis of histologic, serologic, and clinical data only into two groups: as seminomas and nonseminomas or NSGCTs. In a chapter dealing with the pathobiology of germ cells it is important not to belittle this simple empirical approach that has proven its clinical usefulness (Peckham et al. 1981; Vugrin et al. 1988). The widespread use of this simple clinical classification attests to its appeal to practicing oncologists. At the same time it reflects the fact that the clinical advances in the diagnosis and therapy of testicular germ cell tumors have far outpaced the slow and relatively tedious progress made in the understanding of the basic biology of these tumors.

## Conclusion

In this overview of the pathobiology of human germ cell neoplasia, I have tried to touch upon some critical issues and discuss them in the context of a historical perspective, relevant experimental animal tumor data, and recent laboratory findings derived from the analysis of human tumor material and tumor cell lines. The understanding of the pathobiology of human germ cell tumors still lags behind the progress made in the treatment of these tumors. Nevertheless, the lessons learned from studying the human germ cell tumors have already had significant applications in medical practice and are important for the understanding of the basic pathobiology of tumors as well as normal gonadal biology and embryogenesis.

*Acknowledgements.* The secretarial help of Miss Rochelle Hudson is gratefully acknowledged. The original work cited in this article was supported by USPHS grants from the National Institutes of Health, Bethesda, Maryland.

## References


Andrews PW (1988) Human teratocarcinomas. Biochim Biophys Acta 948:17–36

Billmire DF, Grossfeld JL (1986) Teratomas in childhood. Analysis of 142 cases. J Pediatr Surg 21:548–551

Carritt B, Parrington J, Welch HM, Povey S (1982) Diverse origins of multiple ovarian teratomas in a single individual. Proc Natl Acad Sci USA 79:7400–7404

Castedo SMJ, Dejong B, Oosterhuis JW, Seruca R, Idenburg VJS, Dam A, Te Meerman G, Koops HS, Sleijfer DT (1989a) Chromosomal changes in human primary testicular nonseminomatous germ cell tumors. Cancer Res 49:5696–5701

Castedo SMMJ, Dejong B, Oosterhuis JW, Seruca R, Te Meerman G, Dam A, Koops HS (1989b) Cytogenetic analysis of ten human seminomas. Cancer Res 49:439–443

Collins DH, Pugh RCB (1964) Classification and frequency of testicular tumours. Br J Urol [Suppl] 36:1–11

Cortes D, Thorup J, Graem N (1989) Bilateral prepubertal carcinoma in situ of the testis and ambiguous external genitalia. J Urol 142:1065–1068

Crick F (1988) What mad pursuit. Basic, New York

Dal Cin P, Drochmans A, Moerman P, van den Berghe H (1989) Isochromosome 12p in mediastinal germ cell tumor. Cancer Genet Cytogenet 42:243–251

Damjanov, I (1989) Is seminoma a relative or a precursor of embryonal carcinoma? Lab Invest 60:1–3

Damjanov I, Klauber G (1980) Microscopic gonadoblastoma in a dysgenetic gonad of an infant. Urology 15:605–609

Damjanov I, Solter D (1974) Experimental teratoma. Curr Top Pathol 59:69–129

Damjanov I, Drobnjak P, Grizelj V (1975) Ultrastructure of gonadoblastoma. Arch Pathol 99:25–28

Damjanov I, Fox N, Knowles BB, Solter D, Lange PH, Fraley EE (1982) Immunohistochemical localization of murine stage-specific embryonic antigen in human testicular germ cell tumors. Am J Pathol 108:225–230

Dexeus FH, Logothetis CJ, Chong C, Sella A, Ogden S (1988) Genetic abnormalities in men with germ cell tumors. J Urol 140:80–84

Dixon FJ Jr, Moore RA (1953) Testicular tumors. A clinico-pathologic study. Cancer 1953 6:427–454

Eppig JJ (1978) Granulosa cell deficient follicles. Occurrence, structure and relationship to ovarian teratocarcinogenesis in strain LT/Sv mice. Differentiation 12: 111–120

Eppig JJ, Kozak L, Eichner E, Stevens LC (1977) Ovarian teratomas in mice are derived from oocytes that have completed the first meiotic division. Nature 269: 517–520

Ewing J (1911) Teratoma testis and its derivatives. Surg Gynecol Obstet 12:230–261

Falin LI (1969) The development of genital glands and the origin of germ cells in human embryogenesis. Acta Anat (Basel) 72:195–232

Floyd C, Ayala AG, Logothetis CJ, Silva EG (1988) Spermatocytic seminoma with associated sarcoma of the testis. Cancer 61:409–414

Fosså SD, Aass N, Ous S, Hoie J, Stenwig AE, Lien HH, Paus E, Kaalhus O (1989) Histology of tumor residuals following chemotherapy in patients with advanced nonseminomatous testicular cancer. J Urol 140:1239–1242

Fox N, Damjanov I, Martinez-Hernandez A, Knowles BA, Solter D (1981) Immunohistochemical localization of the early embryonic antigen (SSEA-1) in postimplantation mouse embryos and fetal and adult tissues. Dev Biol 83:391–398
Francavilla S, Zamboni L (1985) Differentiation of mouse ectopic germinal cells in intra- and perigonadal locations. J Exp Zool 233:101–109
Friedman NB, Moore RA (1946) Tumors of the testis. A report of 922 cases. Milit Surg 99:573–593
Gardner RL (1983) Origin and differentiation of extraembryonic tissues in the mouse. Int Rev Exp Pathol 24:63–133
Giwercman A, Müller J, Skakkebaek N (1988) Carcinoma in situ of the undescended testis. Semin Urol 6:110–119
Gustavson KH, Rune C (1988) Familial ovarian dermoid cysts. Ups J Med 93:53–56
Hartley AL, Birch JM, Kelsey AM, Marsden HB, Harris M, Teare MD (1989) Are germ cell tumors part of the Li-Fraumeni cancer family syndrome? Cancer Genet Cytogenet 42:221–226
Javadpour N (1986) Misconceptions and source of errors in interpretation of cellular and serum markers in testicular cancer. J Urol 135:879–890
Johnson LC (1953) A general theory of bone tumors. Bull NY Acad Med 29:164–171
Kaplan GW, Cromie WC, Kelalis PP, Silber I, Tank ES Jr (1988) Prepubertal yolk sac testicular tumors – report of the testicular tumor registry. J Urol 140:1109–1112
Kaufman MH (1983) Early mammalian development: parthenogenetic studies. Cambridge University press, New York
Kingsbury AC, Frost F, Cookson WO (1987) Dysgerminoma, gonadoblastoma and testicular germ cell neoplasia in phenotypically female and male siblings with 46 XY genotype. Cancer 59:28–29
Lachman MF, Kim K, Loo B-C (1986) Mediastinal teratoma associated with Klinefelter's syndrome. Arch Pathol Lab Med 110:1067–1071
Mahour GH, Wooley MM, Trivedi SN, Landing BH (1974) Teratomas in infants and childhood: experience with 81 cases. Surgery 76:309–318
Malagolowkin M, Ortega JA, Krailo M, Gonzalez O, Mahour GH, Landing BH, Siegel SE (1989) Immature teratomas: identification of patients at risk for malignant recurrence. JNCI 81:870–874
Manivel JC, Jessurum J, Wick MP, Dehner LP (1987) Placental alkaline phosphatase immunoreactivity in testicular germ cell neoplasms. Am J Surg Pathol 11:21–29
Manivel JC, Simonton S, Wold LE, Dehner LP (1988) Absence of intratubular germ cell neoplasia in testicular yolk sac tumors in children: a histochemical and immunohistochemical study. Arch Pathol Lab Med 112: 641–645
Manivel JC, Reinberg Y, Niehans GA, Fraley EE (1989) Intratubular germ cell neoplasia in testicular teratomas and epidermoid cysts. Cancer 64:715–720
Mann BD, Sparkes RS, Kern DH, Morton DL (1983) Chromosomal abnormalities of a mediastinal embryonal carcinoma in a patient with 47, XXY Klinefelter syndrome: evidence for the premeiotic origin of a germ cell tumor. Cancer Genet Cytogenet 8:191–196
Mostofi FK, Sesterhenn IA, Davis CJ Jr (1987) Immunopathology of germ cell tumors of the testis. Semin Diagn Pathol 4:320–341
Müller JM, Skakkebaek NE, Nielsen OH (1984) Cryptorchidism and testis cancer: atypical infantile germ cells followed by carcinoma in situ and invasive carcinoma in adulthood. Cancer 54:629–634
Mutter GL (1987) Teratoma genetics and stem cells: a review. Obstet Gynecol Rev 42:661–670
Nadimpalli VR, Reyes H, Manaligold JR (1989) Retroperitoneal teratoma with fetuses. Teratology 39:233–236
Ohama K, Nomura K, Okamoto E (1985) Origin of immature teratoma of the ovary. Am J Obstet Gynecol 152:896–900

Oosterhuis JW, Suurmeijer AJH, Sleijfer DT, Koops HS, Oldhoff J, Fleuren GJ (1983) Effects of multiple-drug chemotherapy (*cis*-diammine-dichloroplatinum, bleomycin and vinblastine) on the maturation of retroperitoneal lymph node metastases of non-seminomatous germ cell tumors of the testis. Cancer 51:408–416
Oosterhuis JW, Castedo SMMJ, Dejong B, Cornelisse CJ, Dam A, Sleijfer DT, Koops HS (1989) Ploidy of primary germ cell tumors of the testis: pathogenesis and clinical relevance. Lab Invest 50:14–21
Parrington JM, West LF, Povey S (1984) The origin of ovarian teratomas. J Med Genet 21:4–12
Paiva J, Damjanov I, Lange PH, Harris H (1983) Immunohistochemical localization of placenta-like alkaline phosphatase in testis and germ-cell tumors using monoclonal antibodies. Am J Pathol 111:156–165
Peckham MJ, Barrett A, Mcelwain TJ, Hendry WF, Raghavan D (1981) Non-seminoma germ cell tumours (malignant teratoma) of testis. Results of treatment and an analysis of prognostic factors. Br J Urol 53:162
Pierce GB, Abell MR (1970) Embryonal carcinoma of the testis. Pathol Annu 5:27–60
Pitot HC, Beer D, Hendrich S (1988) Multi-stage carcinogenesis: the phenomenon underlying the theories. In: Iversen O (ed) Theories of carcinogenesis. Hemisphere, New York, pp 159–177
Plachot M, de Grouchy J, Junca A-M, Mandelbaum J, Salat-Baroux J, Cohen J (1988) Chromosome analysis of human oocytes and embryos: does delayed fertilization increase chromosome imbalance? Hum Reprod 3:125–127
Pugh RCB (1976) Pathology of the testis. Blackwell, Oxford
Robboy SJ, Miller T, Donahoe PK, Jahre C, Welch WR, Haseltine FP, Miller WA, Atkins L, Crawford JD (1982) Dysgenesis of testicular and streak gonads in the syndrome of mixed gonadal dysgenesis: perspective derived from a clinico-pathologic analysis of twenty-one cases. Hum Pathol 13:700–720
Rutgers JL, Young RH, Scully RE (1988) The testicular "tumor" of the adrenogenital syndrome. A report of six cases and review of the literature on testicular masses in patients with adrenocortical disorders. Am J Surg Pathol 12:503–513
Shevinsky LH, Knowles BB, Damjanov I, Solter D (1982) Monoclonal antibody to murine embryos defines stage-specific embryonic antigen expressed in mouse embryos and human teratocarcinoma cells. Cell 30:697–705
Silver LM, Martin GR, Strickland S (eds) (1983) Teratocarcinoma stem cells. Cold Spring Harbor Laboratory, Cold Spring Harbor, New York
Skakkebaek NE, Berthelsen JG, Giwercman A, Müller J (1987) Carcinoma-in-situ of the testis: possible origin from gonocytes and precursor of all types of germ cell tumours except spermatocytoma. Int J Androl 10:19–28
Solter D (1988) Differential imprinting and expression of maternal and paternal genomes. Annu Rev Genet 22:127–146
Solter D, Damjanov I (1979) Teratocarcinoma and the expression of oncodevelopmental genes. Methods Cancer Res 18:277–332
Solter D, Skreb N, Damjanov I (1970) Extrauterine growth of mouse egg-cylinders results in malignant teratoma. Nature 1970 227:503–504
Srigley JR, Mackay B, Toth P, Ayala A (1988) The ultrastrucuture and histogenesis of male germ neoplasia with emphasis on seminoma with early carcinomatous features. Ultrastruct Pathol 12:67–86
Stephens TD, Spall R, Urfer AG, Martin R (1989) Fetus amorphus or placental teratoma? Teratology 40:1–10
Stevens LC (1967) The biology of teratomas. Adv Morphog 6:1–31
Stevens LC (1970a) The development of transplantable teratocarcinomas from intratesticular grafts of pre- and post-implantation mouse embryos. Dev Biol 21:364–381
Stevens LC (1970b) Experimental production of testicular teratomas in mice of strains 129, A/He and their $F_1$ hybrids. JNCI 44:929–932

Stevens LC, Varnum D (1974) The development of teratomas from parthenogenetically activated ovarian eggs. Dev Biol 37:369–380
Teshima S, Shimosato Y, Hirohashi S, Tome Y, Hayashi I, Kanazawa H, Kakizoe T (1988) Four new human germ cell tumor cell lines. Lab Invest 59:328–336
True LD, Otis CN, Dezpardo W, Scully RE, Rosai J (1988) Spermatocytic seminoma of testis with sarcomatous transformation. A report of five cases. Am J Surg Pathol 12:75–82
Ulbright TM, Loehrer PJ (1988) Choriocarcinoma-like lesions in patients with testicular germ cell tumors. Two histologic variants. Am J Sug Pathol 12:531–541
Ulbright TM, Roth LM, Brodhecker, CA (1986) Yolk sac differentiation in germ cell tumors. A morphologic study of 50 cases with emphasis on hepatic, enteric and parietal yolk sac features. Am J Surg Pathol 10:151–164
Upadhyay S, Zamboni L (1982) Ectopic germ cells. Natural model for the study of germ cell sexual differentiation. Proc Natl Acad Sci USA 79:6584–6588
Vandeputte M, Sobis H (1988) Experimental rat model for human yolk sac tumor. Eur J Cancer Clin Oncol 24:551–558
Vugrin D, Chen A, Feigl P, Laszlo J (1988) Embryonal carcinoma of the testis. Cancer 61:2348–2352
Wilms M (1896) Die teratoiden Geschwulste des Hodens mit Einschluβ der sog. Cystoide und Enchondrome. Beitr Pathol Anat 19:233–366
Wilms M (1898) Embryome und embryoide Tumoren des Hodens. Dtsch Z Chir 49:1–12
Zamboni L, Upadhyay S (1983) Germ cell differentiation in mouse adrenal glands. J Exp Zool 228:173–193

# Cells of Origin

## *Carcinoma in Situ of the Testis: Possible Origin, Clinical Significance, and Diagnostic Methods*

A. Giwercman, J. Müller, and N.E. Skakkebæk

University Department of Growth and Reproduction GR, Rigshospitalet, Section 4052, Blegdamsvej 9, 2100 Copenhagen, Denmark

The pathogenesis of testicular germ cell tumors is still a controversial topic. Thus, it is unclear whether nonseminomatous cancer is preceded by a seminoma stage (Friedman, 1951; Raghavan et al. 1982; Oosterhuis et al. 1989) or whether these two types develop directly from a precursor cell (Pierce and Abell 1970; Skakkebæk and Berthelsen 1981). However, both pathogenetic models include the stage of carcinoma in situ (CIS) (Skakkebæk 1972) as a precursor of all testicular germ cell tumors, with spermatocytic seminoma as the only exception (Müller et al. 1987).

The association between the CIS pattern and subsequent development of invasive cancer was not recognized until the early 1970s (Skakkebæk 1972). Since then, an increasing amount of information on various aspects of CIS has been accumulated. In this chapter we will focus on data indicating that CIS germ cells are in fact malignant gonocytes. Additionally, the clinical significance and methods of diagnosing CIS will be discussed with most emphasis on recently reported methods of detection of malignant germ cells in seminal fluid.

### Origin of CIS Germ Cells

There is increasing evidence that CIS is an inborn lesion, probably arising in early fetal life (Skakkebæk et al. 1987). Thus, CIS cells were identified in the testes of several prepubertal individuals, in one as early as 1 month of age (Müller et al. 1985). The prepubertal CIS pattern found in a maldescended testis of a 10-year-old boy was followed by a postpubertal CIS found in a biopsy performed at the age of 20 years. Subsequently this patient developed invasive testicular cancer (Müller et al. 1984).

Additionally, epidemiological studies indicate a common etiological factor of testicular neoplasia and cryptorchidism (Swerdlow et al. 1983; Giwercman et al. 1988a), which is in favor of the prenatal origin of germ cell tumors.

Recent Results in Cancer Research, Vol. 123

Oosterhuis et al. (1989) suggested, on the basis of microspectrophotometric (Müller et al. 1981) and flow cytometric DNA measurements, polyploidization as an initial event in the pathogenesis of testicular germ cell malignancy. Recently, the hypothesis that CIS germ cells are malignant gonocytes was proposed (Skakkebæk et al. 1987), based mainly on data showing morphological, histochemical, and immunological similarities between these two cell types.

However, the factors which might trigger malignant transformation of fetal gonocytes into CIS cells are still not known. In Denmark, men born during the two World Wars exhibit a lower testicular cancer risk than those in birth cohorts preceding and following the war (Møller 1989). This finding indicates an environmental factor being of etiological importance for the development of testicular cancer. Endocrine disturbances, including a relative excess of estrogen and perhaps progesterone at the time of testicular differentiation, have also been proposed as a major risk factor for neoplasia of the testis (Henderson et al. 1979).

## *Morphological Aspects*

Light microscopic examination reveals that CIS germ cells resemble gonocytes from fetal gonads. Thus, the median nuclear diameter of CIS cells is 9.7 μm (range, 9.2–10.5 μm), which is significantly higher than the 6.5 μm (range, 5.7–7.1 μm) of spermatogonia (Müller 1987). Also the ultrastructure of CIS germ cells bears some resemblance to that of primitive germ cells (Nielsen et al. 1974; Gondos et al. 1983; Holstein et al. 1987). Nuage bodies and dense-cored vesicles previously used for identification of primordial germ cells in rodents have also been described in CIS germ cells. The finding of several irregular nucleoli gives further support to the concept that CIS germ cells are of primitive origin.

In typical cases of postpubertal CIS (Skakkebæk 1972, 1978) (Fig. 1), the tubules with CIS cells are composed of a single row of CIS germ cells located between normally appearing Sertoli's cells and the thickened basement membrane. Rarely, CIS germ cells and normal germ cells including spermatogonia, spermatocytes, and spermatids may be found within the same tubular cross section. Such a pattern represents a transition zone between CIS and normal seminiferous epithelium. Lymphocyte infiltrates may sometimes be found in the interstitial tissue surrounding the tubules harboring CIS cells.

In prepubertal testes (Müller et al. 1984), the distribution of CIS cells is somewhat different from the postpubertal pattern as the cells are not located in a single row but rather are irregularly distributed throughout the seminiferous tubules. According to the above-mentioned resemblance between CIS cells and gonocytes, the former may, in routine histological preparation, be somewhat difficult to distinguish from normal, prepubertal germ cells (Müller 1987).

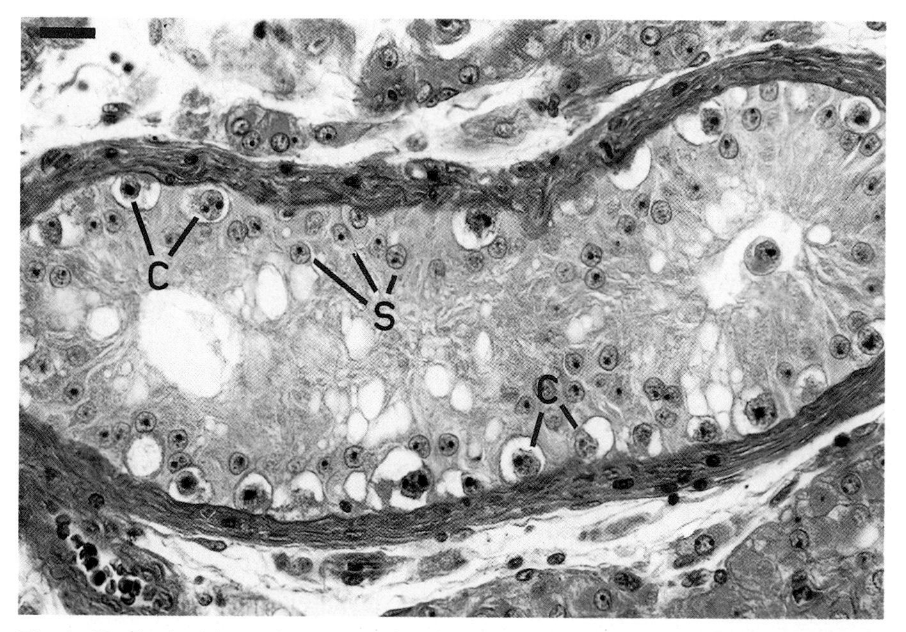


**Fig. 1.** Testicular biopsy from a previously maldescended testis of a 26-year-old man. Characteristic CIS pattern with CIS cells (*C*), normally appearing Sertoli's cells (*S*), and a thickened basement membrane. Stieve's fixative. Iron H & E, *bar*, 25 μm

### *Histochemical and Immunological Aspects*

Accumulations of glycogen, which is considered a characteristic marker of primitive fetal germ cells in mammals, can also be demonstrated in the cytoplasm of CIS germ cells (Holstein et al. 1987; Koide et al. 1987). Another common histochemical characteristic is the alkaline phosphatase content in very early fetal gonocytes (McKay et al. 1953) as well as in CIS cells (Beckstead 1983). These histochemical findings have been confirmed by immunohistochemical studies with antibodies against placental-like alkaline phosphatase (PlAP). Thus, CIS germ cells were shown to be PlAP positive (Jacobsen and Nørgaard-Pedersen 1984; Manivel et al. 1987) and cells containing PlAP were also demonstrated alongside the genital ridge of a 7.5 mm crown-rump embryo (Hustin et al. 1987). Finally, like the fetal testicular germ cells, CIS cells express the surface marker defined by the monoclonal antibody to the murine stage-specific embryonic antigen 3 (SSEA-3) (Damjanov 1986).

## Progression from CIS to Invasive Cancer

It still remains to be elucidated whether CIS cells have the capacity to progress directly into the different types of malignant germ cell tumors (Pierce and Abell 1970; Skakkebæk et al. 1987) or whether the nonseminomas are preceded by the stage of seminoma (Friedman 1951; Raghavan et al. 1982; Oosterhuis et al. 1989).

At the microscopic level the malignant potential of CIS can be demonstrated by the invasive growth of CIS cells in the interstitial tissue or rete testis. However, it has been postulated that a "second hit" is necessary for transformation of CIS cells into an invasive neoplasm (Damjanov 1986). At present it is not known what represents such a "second hit." Endocrine factors may play a role in the development of invasive cancer of the testis, as the age of peak incidence of testicular cancer is related to the occurrence of the postnatal and the postpubertal increase in gonadotropin and/or testosterone production (Skakkebæk et al. 1987).

Despite the fact that several years may pass from the time of diagnosis of CIS to the development of a tumor, our clinical experience indicates that CIS in most, if not all, cases progresses into an invasive neoplasm. Regression of CIS has not been observed in testes of men who had serial biopsies performed. Additionally, recently we found the overall prevalence of testicular neoplasia, including CIS, in testes of men aged 18–50 years who suffered sudden unexpected death to be of the same magnitude as the lifetime risk of testicular cancer in the Danish male population (Giwercman et al. 1990c). Thus, it seems that regression of CIS or persistence at the noninvasive stage is a very rare or a nonoccurring event.

## Clinical Significance of CIS

Actuarial "life-table" analyses have demonstrated that 70% of men with CIS of the testis will develop an invasive cancer within 7 years of the first biopsy. This figure is based on observation of infertile men (Skakkebæk et al. 1982) as well as men treated for a unilateral testicular cancer and harboring CIS in the contralateral testis (von der Maase et al. 1986a). As spontaneous regression of CIS has not yet been observed, we believe that the remaining 30% of men, if still untreated, would sooner or later develop a tumor. Thus, in our opinion, the surveillance which was recommended in the 1970s, when the spontaneous course of CIS was still unknown, should for ethical reasons be abandoned in favor of active management (see below).

Much knowledge has now been accumulated on the prevalence of CIS in – mainly postpubertal – individuals, including both the general population and different high-risk groups of testicular neoplasia.

### *General Male Population*

No data on the prevalence of CIS in the gonads of healthy, young males are available. In a study of autopsy material, no CIS was found in testicular specimens from 399 men aged 18–50 years who suffered sudden unexpected death (Giwercman et al. 1990c). However, two of these men had previously been treated for testicular tumor and an additional individual had had one gonad removed due to CIS. Thus, the overall prevalence of testicular malignancy including CIS was 0.8%. This figure is very close to the 0.7% lifetime risk of testicular cancer in the Danish male population (Danish Cancer Registry 1988).

### *Contralateral Testis in Men with Unilateral Cancer of the Testis*

Among 600 Danish men treated for a unilateral testicular cancer, 34 were found to have CIS in the contralateral testis. This corresponds to a prevalence of 5%–6% (Berthelsen et al. 1982; von der Maase et al. 1987), which agrees well with the frequency of bilateral testicular cancer found in other countries (Scheiber et al. 1987; Thompson et al. 1988).

### *Cryptorchidism*

Men with a history of cryptorchidism are known to have an increased risk of testicular cancer. Although some studies stated the increased relative risk to be 40- to 50-fold (Gilbert and Hamilton 1940; Campbell 1942), more recent data indicate that these patients have a 4- to 10-fold increased risk of testicular cancer (Morrison 1976; Henderson et al. 1979; Schottenfeld et al. 1980; Pottern et al. 1985; Giwercman et al. 1987). In agreement with these data, the combined results of 3 Scandinavian studies (Krabbe et al. 1979; Pedersen et al. 1987; Giwercman et al. 1989) based on testicular biopsies from a total of 444 men with previous maldescent indicate that the prevalence of CIS in this category of men in 2%–3%, which is approximately 4 times higher than the figure found in autopsy material (Giwercman et al. 1990c).

Most studies on CIS in maldescended testes have been carried out on postpubertal individuals. CIS has also been diagnosed in prepubertal maldescended testes with subsequent progression to an adult CIS pattern and tumor development (Müller et al. 1984). In two relatively small series of testicular biopsies from maldescended, prepubertal gonads, no case of CIS was found (Muffly et al. 1984; Müller and Skakkebæk 1984a). However, it must be assumed that the number of CIS germ cells is lower in prepubertal than in postpubertal testes, which may imply difficulties in diagnosing this premalignant condition before puberty (Müller 1987).

### *Infertile Men*

As yet, no systematic screening studies on the prevalence of CIS in infertile men have been reported. In retrospective studies of testicular specimens from

Swiss (Nüesch-Bachmann and Hedinger 1977), Danish (Skakkebæk 1978), and British (Pryor et al. 1983) men, CIS was detected in the gonads of 0.4%–1.1% of the individuals. This rather inhomogeneous material included, however, males with an *a priori* known low risk of testicular cancer, such as men with Klinefelter's syndrome and Sertoli-cell-only syndrome. Based on the experience from other risk groups, the highest frequency of CIS would be expected among infertile men with severe oligozoospermia, atrophic testes, or a history of maldescent (Berthelsen et al. 1982). In a recent German study, CIS was found in the testes of approximately 1% of men with a sperm count below 20 million/ml (Schütte 1988).

### *Intersex Conditions*

Intersex patients, although rather rare, have an extremely high risk of testicular malignancy if their karyotype includes a Y chromosome (Manuel et al. 1976). In keeping with this, Müller et al. (1985) found CIS in the gonads of all four individuals with gonadal dysgenesis and a 45,X0/46,XY mosaic karyotype in their study. Additionally, 4 of 12 patients with the androgen insensitivity syndrome had CIS (Müller and Skakkebæk 1984b).

### *Extragonadal Germ Cell Tumors*

Daugaard et al. (1987) found 8 patients with CIS of the testis among 15 men with an assumed extragonadal germ-cell tumor, none of whom had clinical signs of testicular tumors. Interestingly, CIS of the testis was only found in patients with the extragonadal component located to the retroperitoneum, not in men with a mediastinal tumor. It remains to be elucidated whether the extragonadal germ cell tumors are caused by an aberrant migration of germ cells during the early fetal life or whether they represent metastases of testicular origin.

## Management

The primary goal for the management of CIS of the testis is to prevent the development of an invasive tumor and, if possible, simultaneous preservation of sufficient endogenous androgen production. In view of the data on the spontaneous course of CIS (Skakkebæk et al. 1982; von der Maase et al. 1986a), therapeutic intervention is recommended in adults with CIS. In prepubertal individuals, management depends on the patient category.

### *Adults*

In men with unilateral CIS and no sign of neoplasia in the other gonad, unilateral orchiectomy is the treatment of choice. This treatment usually does not affect the serum testosterone level. Also the semen quality will remain unchanged as the testis harboring the CIS usually has very poor spermatogenesis.

If both testes are affected by CIS or, more commonly, if one testis has been removed because of cancer and the contralateral testicle harbors CIS, localized irradiation is recommended. Twenty men of the latter category, who received a dose of 20 Gray delivered as 10 fractions, have now been followed for up to 5 years (von der Maase et al. 1986b, 1987). This treatment appears completely to eradicate CIS, as the CIS pattern in all cases was converted into Sertoli-cell-only syndrome without the presence of any germ cells. Despite some injury of Leydig's cells caused by the radiation treatment, as indicated by an increase in serum luteinizing hormone (LH) levels, no significant changes in testosterone concentration were observed. However, long-term follow-up is necessary in order to exclude that: (a) some CIS cells "survived" radiation treatment and (b) the adverse effect on Leydig's cells is of clinical significance for the patients.

Localized irradiation is not recommended in men with unilateral CIS and no neoplasia in the other testis since coirradiation of the contralateral testicle might induce genetic damage to the germ cells. However, if long-term follow-up confirms that CIS cells are completely eradicated by irradiation, this treatment might be an alternative if the testis without CIS has no sperm production.

It is still an open question whether radiotherapy for CIS should be offered to men who are receiving chemotherapy due to a metastatic testicular cancer. CIS cells were observed to disappear from testicular biopsies in men treated with the Einhorn regime (von der Maase et al. 1985). However, recent reports indicate that, at least in some cases, cytotoxic drugs rather reduce the number of CIS cells but do not completely eradicate them (von der Maase et al. 1988). Accordingly, testicular cancer in the contralateral testis has previously been reported in men who had received Einhorn treatment (Fowler et al. 1979; Fosså and Aass 1989).

### *Children*

The spontaneous course of CIS diagnosed in prepubertal individuals has not yet been thoroughly investigated. Generally, a close follow-up until the diagnosis can be confirmed in a subsequent biopsy performed after puberty is therefore recommended. An exception may be intersex patients. In view of the extremely high risk of gonadal tumors, an orchiectomy, when the diagnosis is established, may be recommended in these individuals. However, in each of these rare cases an individual assessment is needed. The role of irradiation in the treatment of prepubertal CIS has not been studied. A limiting factor might be the fact that Leydig's cells are more radiosensitive in prepubertal individuals than in adults (Shalet et al. 1978).

## Methods of Diagnosis

At present, surgical, testicular biopsy is the only established method of diagnosing CIS of the testis. In adults, CIS seems to be a disperse lesion, and testicular biopsy performed after puberty was shown to be a very sensitive tool in the detection of this early neoplasia (Berthelsen and Skakkebæk 1981). Testicular biopsy can be performed with the patient under local anesthesia as an outpatient procedure, which is well accepted by patients and associated with complications in only very few cases (Rowley and Heller 1966; Bruun et al. 1987).

It is possible to diagnose CIS by light microscopic examination of routinely stained histological sections, particularly if Bouin's or Stieve's fluid is used as a fixative instead of formaldehyde, as the latter destroys the testicular morphology. However, even in properly treated sections CIS may be overlooked (Pryor et al. 1983; Nistal et al. 1989). The diagnosis of CIS may be facilitated by using immunohistochemical techniques.

### *Immunohistochemical Markers of CIS Cells*

In our experience three different immunohistochemical markers have proved to be of value in the detection of CIS germ cells. *PlAP* has for several years been known as a rather sensitive marker of CIS (Jacobsen and Nørgaard-Pedersen 1984; Manivel et al. 1987). Recently, two monoclonal antibodies (MAbs), *M2A* and *43–9F*, were also shown to bind to CIS germ cells in an immunohistochemical staining procedure (Fig. 2). MAb M2A was raised against a human ovarian adenocarcinoma cell line HEY and reacted against a still uncharacterized carbohydrate epitope on the surface of CIS cells (Giwercman et al. 1988b) as well as cells of seminomas and dysgerminomas, but not nonseminomas (Bailey et al. 1986). MAb 43–9F, an antibody against a human squamous cell lung carcinoma line RH-SLC-L11 (Olsson et al. 1984), recognizes an $Le^{a}$-X oligosaccharide determinant on the cell membrane of CIS cells (Mårtensson et al. 1988; Giwercman et al. 1990a). Unlike M2A, 43–9F seems to react strongly with cells of embryonal carcinoma rather than nonseminomas (Giwercman et al. 1990a). MAb 43–9F was shown also to be of value in the detection of prepubertal CIS. As the MAb M2A also binds to immature Sertoli's cells (Baumal et al. 1989), this MAb is not suitable for the diagnosis of prepubertal CIS.

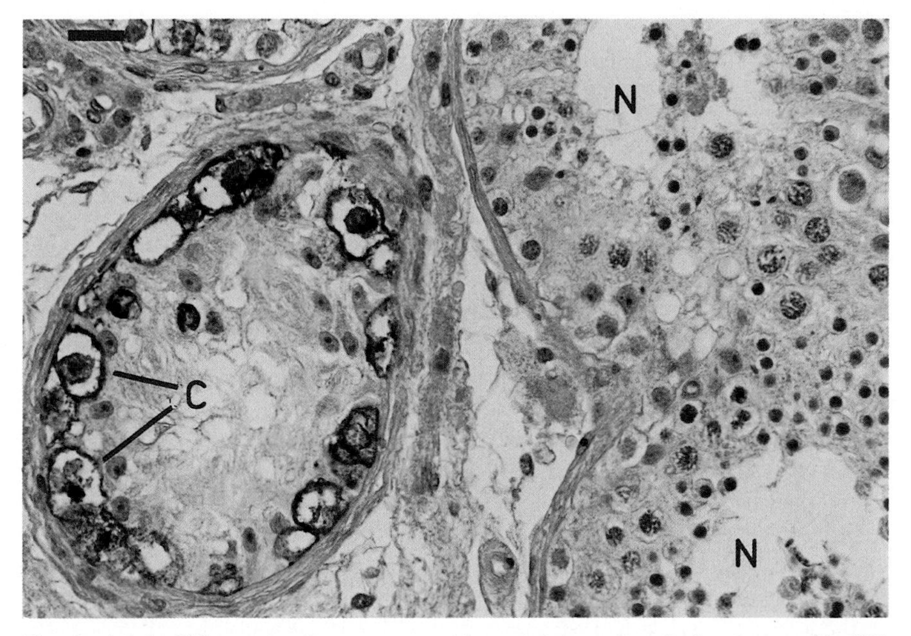


**Fig. 2.** Adult CIS pattern. Immunoperoxidase staining of testicular tissue with CIS with the monoclonal antibody 43–9F. Streptavidine-biotin method. The color reaction was developed with 3.3′-diaminobenzidine. Note the color reaction on the surface of CIS cells (*C*). No reaction in tubules with ongoing spermatogenesis without CIS (*N*). Stieve's fixative, *bar*, 25 μm

### *Testicular Biopsy as Screening for CIS*

This procedure is recommended as screening for CIS in some selected groups of patients with a high risk of CIS, i.e., intersex patients or in the contralateral testis of men treated for a unilateral cancer. Biopsy should also be considered in men with a history of cryptorchidism, whereas the data on the prevalence of CIS in infertile men are not yet sufficient to justify routine biopsy in these men. Most cases of testicular cancer are, however, detected in individuals outside the above-mentioned high-risk groups. Alternative methods for diagnosis of CIS of the testis are therefore currently being tested.

### *Other Invasive Methods*

Diagnosis of CIS might be simplified by the use of a needle biopsy. By means of DNA flow cytometry, aneuploid cells were identified in a fine needle aspirate from a testis harboring CIS (Nagler et al. 1990). However, preliminary data indicate that the diagnostic sensitivity of TRU-CUT biopsy or

DNA flow cytometry of testicular fine-needle aspirates may not be sufficiently high for screening purposes (Clausen et al. 1990; unpublished data from our laboratory).

### *Noninvasive Methods of Screening*

Carcinoma in situ germ cells are like other germ cells of the seminiferous epithelium located inside the seminiferous tubules. Cells such as spermatogonia, spermatocytes, and spermatids are known to be shed in the semen (WHO 1987). One would, therefore, expect to find malignant germ cells in ejaculates of men with CIS. Morphologically abnormal cells were detected in the semen of patients with testicular cancer (Czaplicki et al. 1987). Generally, however, the morphology of seminal cells other than sperm cells is poorly preserved. Attempts have therefore been made to identify CIS cells in semen using previously identified markers such as MAb M2A and aneuploidy (Müller et al. 1981).

The first evidence of the presence of CIS cells in semen originated from DNA flow cytometric studies which revealed the presence of aneuploid cells in the seminal fluid of four of eight men with CIS. Aneuploid cells could no longer be detected in seminal fluid 1 year after these four men were treated for CIS – additional evidence indicating that the aneuploid cells were in fact CIS cells (Giwercman et al. 1988c). However, the sensitivity of DNA flow cytometry in the detection of CIS in semen is apparently not sufficiently high, and other methods of identification of malignant germ cells in semen are now being tested.

Thus, cells reacting with the MAb M2A were found in semen from men with CIS (Giwercman et al. 1988d). This study has also shown that the number of CIS cells shed in semen may be too low to detect by flow cytometry. A lack of specificity, due to an aberrant staining of nonviable cells present in the seminal fluid, has as yet limited the use of immunocytochemistry in the screening for CIS.

DNA in situ hybridization, a method which recently has been shown to be more sensitive than flow cytometry in the detection of aneuploid cells (Hopman et al. 1988), was therefore used on seminal smears from men with CIS and from controls (Giwercman et al. 1990b). Biotinylated DNA probe *pUC 1.77* (Cooke and Hindley 1979), specific for chromosome 1, was used in the in situ hybridization procedure. In paraffin sections, a significant proportion of CIS cells were recently shown to possess an increased (>2) copy number of chromosome 1 (Walt et al. 1989) (Fig. 3a). Preliminary studies have demonstrated that, compared with controls, a significantly higher proportion of cells in ejaculates from men with CIS have three or more copies of chromosome 1 (Fig. 3b). Additionally, out of specimens from 2 men with CIS, 6 men with CIS in a vicinity of a tumor, and 12 controls, using in situ hybridization, it was possible blindly to identify ejaculates from both men

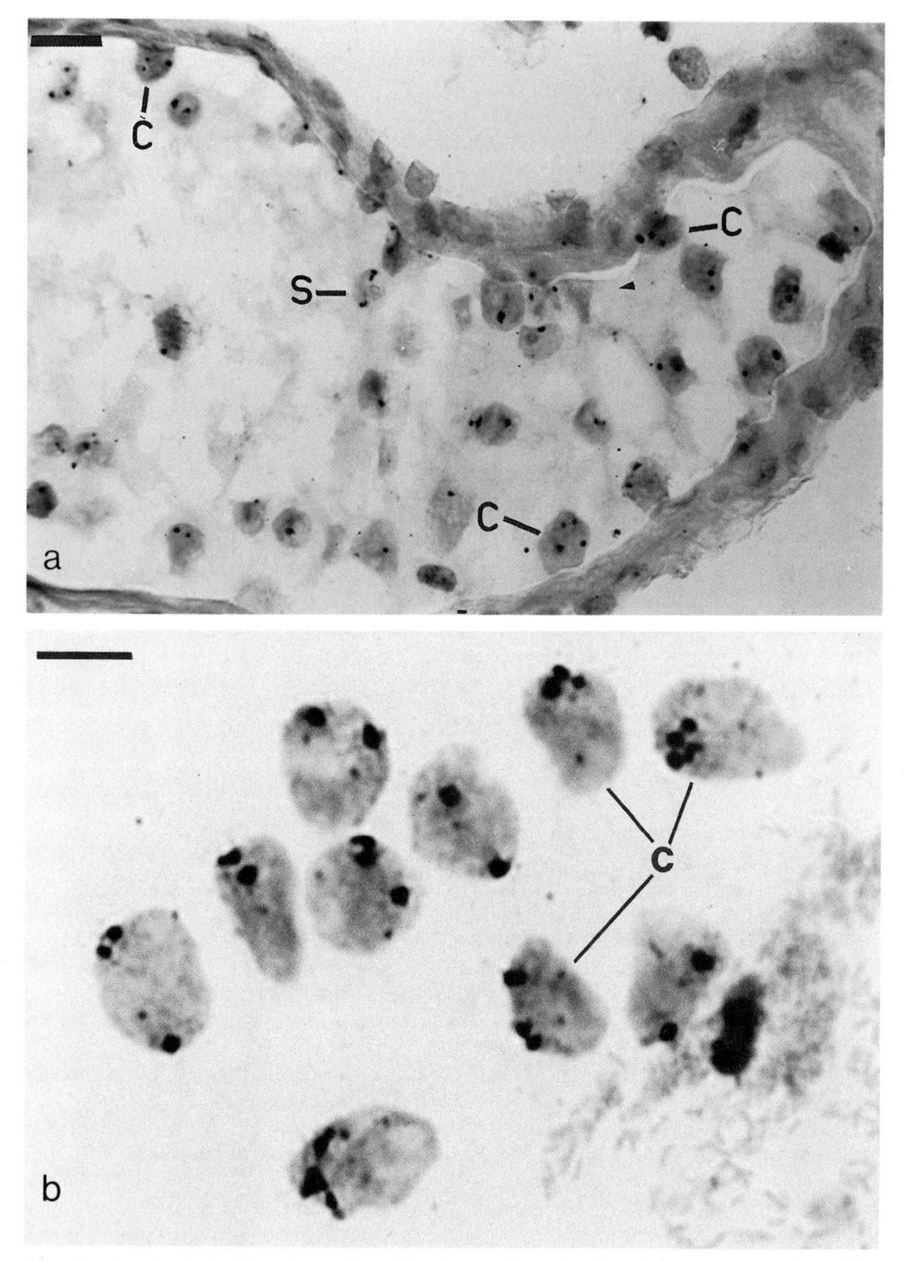


**Fig. 3.a** In situ hybridization with a biotinylated probe for chromosome 1. Paraffin-embedded, formaldehyde-fixed tissue with CIS. Each *dark spot* in the nucleus represents a copy of chromosome 1. Note the increased copy number of chromosome 1 in CIS cells (*C*) but not in Sertoli's cells (*S*), *bar*, 25 μm. **b** In situ hybridization used on a seminal smear from a patient with CIS. Same method as in **a** Cells with an increased copy number of chromosome 1 (*C*) were identified. *Bar*, 10 μm. (From Giwercman et al. 1990b)

with isolated CIS (Giwercman et al. 1990b). Additionally, samples from three of six men, in whom CIS was associated with a tumor. Thus, in situ hybridization with chromosome-specific markers may, alone or in combination with immunohistochemistry, become a valuable tool in screening for CIS of the testis.

Finally, ultrasound scanning may contribute to identification of men with CIS (Lenz et al. 1987).

## Concluding Remarks

The data on CIS of the testis collected through the past 20 years have given new information on the etiology and pathogenesis of testicular germ cell cancer. It now seems obvious that seminomatous as well as nonseminomatous tumors are preceded by CIS, which seems to have its origin in the very early fetal gonocytes. Active intervention in the management of CIS is necessary to prevent the development of testicular cancer. It is realistic to predict that a noninvasive method of screening for CIS can be developed. This will enable us to screen large groups for early testicular neoplasia and thereby efficiently prevent testicular tumors.

*Acknowledgements.* Our studies cited in this review were supported by grants from the Danish Cancer Society and the Daell Foundation.

## References

Bailey D, Baumal R, Law J, Sheldon K, Kannampuzha P, Stratis M, Kahn H, Marks A (1986) Production of monoclonal antibody specific for seminomas and dysgerminomas. Proc Natl Acad Sci USA 83:5291–5295

Baumal R, Bailey D, Giwercman A, Skakkebæk NE, Stratis M, Marks A (1989) A novel maturation marker for human Sertoli cells. Int J Androl 12:354–359

Beckstead J (1983) Alkaline phosphatase histochemistry in human germ cell neoplasms. Am J Surg Pathol 4:341–349

Berthelsen JG, Skakkebæk NE (1981) Value of testicular biopsy in diagnosing carcinoma in situ testis. Scand J Urol Nephrol 15:163–168

Berthelsen JG, Skakkebæk NE, von der Maase, H, Sørensen BL, Mogensen P (1982) Screening for carcinoma in situ of the contralateral testis in patients with germinal testicular cancer. Br Med J 285:1683–1686

Bruun E, Frimodt-Møller C, Giwercman A, Lenz S, Skakkebæk NE (1987) Testicular biopsy as an outpatient procedure in screening for carcinoma-in-situ: complications and the patient's acceptance. Int J Androl 10:199–202

Campbell HE (1942) Incidence of malignant growth of the undescended testicle: a critical and statistical study. Arch Surg 44:353–369

Clausen OPF, Giwercman A, Jørgensen N, Bruun, E, Frimodt-Møller C, Skakkebæk NE (1990) DNA distributions in maldescended testis: hyperdiploid populations without evidence of germ cell neoplasia. Cytometry

Cooke HJ, Hindley J (1979) Cloning of human satellite III DAN: different components are on different chromosomes. Nucleic Acids Res 10:3177–3197
Czaplicki M, Rojewska J, Pykalo R, Szymanska K (1987) Detection of testicular neoplasms by cytological examination of seminal fluid. J Urol 138:767–768
Damjanov I (1986) Testicular germ cell tumors as model of carcinogenesis and embryogenesis. In: Javad pour N (ed) Principles and management of testicular cancer. Thieme, New York, pp 73–87
Danish Cancer Registry (1988) Cancer incidence in Denmark. Danish Cancer Society, Copenhagen
Daugaard G, von der Maase H, Olsen J, Rørth M, Skakkebæk NE (1987) Carcinomain-situ testis in patients with assumed extragonadal germ cell tumours. Lancet 2:528–530
Fowler JE Jr, Vugrin D, Cvitkovic E, Whitmore WF Jr (1979) Sequential bilateral germ cell tumors of the testis despite interval chemotherapy. J Urol 122:421–425
Fosså SD, Aass N (1989) Cisplatin-based chemotherapy does not eliminate the risk of a second testicular cancer. Br J Urol 63:531–534
Friedman NB (1951) The comparative morphogenesis of extragenital and gonadal teratoid tumors. Cancer 4:265–276
Gilbert JB, Hamilton JB (1940) Studies in malignant testis tumors. III. Incidence and nature of tumors in ectopic testes. Surg Gynecol Obstet 71:731–743
Giwercman A, Grindsted J, Hansen B, Jensen OM, Skakkebæk NE (1987) Testicular cancer risk in boys with maldescended testis: a cohort study. J Urol 138:1214–1216
Giwercman A, Müller, J, Skakkebæk NE (1988a) Cryptorchidism and testicular neoplasia. Horm. Res 30:157–163
Giwercman A, Marks A, Bailey D, Baumal R, Skakkebæk NE (1988b) A monoclonal antibody as a marker for carcinoma-in-situ germ cells of the human adult testis. Apmis 96:667–670
Giwercman A, Clausen OPF, Skakkebæk NE (1988c) Carcinoma-in-situ of the testis: aneuploid cells in semen. Br Med J 296:1762–1764
Giwercman A, Marks A, Skakkebæk NE (1988d) Carcinoma-in-situ germ cells exfoliated from seminiferous epithelium into seminal fluid. Lancet 1:530–531
Giwercman A, Bruun E, Frimodt-Møller, C (1989) Prevalence of carcinoma-in-situ and other histopathological abnormalities in testes with a history of cryptorchidism. J Urol 142:998–1002
Giwercman A, Lindenberg S, Kimber SJ, Andersson T, Müller J, Skakkebæk NE (1990a) Monoclonal antibody 43–9F as a sensitive immunohistochemical marker of carcinoma in situ of human testis. Cancer (in press)
Giwercman A, Hopman AHN, Ramaekers FCS, Skakkebæk NE (1990b) Carcinoma in situ of the testis. Detection of malignant germ cells in seminal fluid by means of in situ hybridization. Am J Pathol 136 (in press)
Giwercman A, Müller J, Skakkebæk NE (1990c) Prevalence of carcinoma-in-situ and other histopathologic abnormalities in testes from 399 men who suffered sudden unexpected death. J Urog
Gondos B, Berthelsen JG, Skakkebæk NE (1983) Intratubular germ cell neoplasia (carcinoma-in-situ): a preinvasive lesion of the testis. Ann ClinLab Sci 13:185–192
Henderson BF, Benton B, Jing J, JU, MC, Pike MC (1979) Risk factors for cancer of the testis in young men. Int J Cancer 23:598–602
Holstein AF, Schütte B, Becker H, Hartmann M (1987) Morphology of normal and malignant germ cells. Int J Androl 10:1–18
Hopman AHN, Ramaekers FCS, Raap AK, Beck JLM, Devilee P, Ploeg, VDM, Vooijs GP (1988) In situ hybridization as a tool to study numerical chromosome aberrations in solid bladder tumors. Histochemistry 89:307–316
Hustin J, Collette J, Franchimont P (1987) Immunohistochemical demonstration of placental alkaline phosphatase in various states of testicular development and in germ cell tumours. Int J Androl 10:29–35

Jacobsen GK, Nørgaard-Pedersen B (1984) Placental alkaline phosphatase in testicular germ cell tumours and in carcinoma-in-situ of the testis. Acta Pathol Microbiol Scand [A] 92:323–329
Koide O, Iwai S, Baba K, Iri H (1987) Identification of testicular atypical cells by an immunohistochemical technique for placental alkaline phosphatase. Cancer 60:1325–1330
Krabbe S, Berthelsen JG, Volsted P, Eldrup J, Skakkebæk NE, Eyben FV, Mauritzen K, Nielsen AH (1979) High incidence of undetected neoplasia in maldescended testes. Lancet 1:999–1000
Lenz S, Giwercman A, Skakkebæk NE, Bruun E, Frimodt-Møller C (1987) Ultrasound in detection of early neoplasia of the testis. Int J Androl 10:187–190
Manivel JC, Jessurun J, Wick MR, Dehner L (1987) Placental alkaline immunoreactivity in testicular germ cell neoplasms. Am J Surg Pathol 11:21–29
Manuel M, Katayama KP, Jones HW (1976) The age of occurrence of gonadal tumors in intersex patients with a Y chromosome. Am J Obstet. Gynecol 124:293–300
Mårtensson S, Due C, Påhlsson P, Nilsson B, Eriksson H, Zopf D, Olsson L, Lundblad A (1988) A carbohydrate epitope associated with human squamous lung cancer. Cancer Res 48:2125–2131
Mckay DG, Hertig AT, Adams EC, Daziger S (1953) Histochemical observations on the germ cells of human embryos, Anat Rec 117:201–219
Morrison AS (1976) Cryptorchidism, hernia, and cancer of the testis. JNG'l 56: 731–736
Muffly KE, McWorter CA, Bartone FF, Gardner PJ (1984) The absence of premalignant changes in the cryptorchid testis before adulthood. J Urol 131:523–524
Müller J (1987) Abnormal infantile germ cells and development of carcinoma-in-situ in maldeveloped testes: a stereological and densitometric study. Int J Androl 10:543–567
Müller J, Skakkebæk NE (1984a) Abnormal germ cells in maldescended testes: a study of cell density, nuclear size and deoxyribonucleic acid content in testicular biopsies from 50 boys. J Urol 131:730–733
Müller J, Skakkebæk NE (1984b) Testicular carcinoma in situ in children with the androgen insensitivity (testicular feminization) syndrome. Br Med J 288:1419–1420
Müller J, Skakkebæk NE, Lundsteen C (1981) Aneuploidy as a marker for carcinoma-in-situ of the testis. Acta Pathol Microbiol Scand [A] 89:67–68
Müller J, Skakkebæk NE, Nielsen OH, Græm N (1984) Cryptorchidism and testis cancer. Atypical infantile germ cell cells followed by carcinoma-in-situ and invasive carcinoma in adulthood. Cancer 54:629–634
Müller J, Skakkebæk NE, Ritzén M, Plöen L, Petersen, K (1985) Carcinoma-in-situ of the testis in children with 45, V/46, XY gonadal dysgenesis. J Pediatr 106:431–436
Müller J, Skakkebæk NE, Parkinson, MC (1987) The spermatocytic seminoma: views on pathogenesis. Int J Androl 10:146–156
Møller H (1989) Decrease of testicular cancer risk in men born in war-time. JNG'l 81:1668–1669
Nagler HM, Kaufman DG, O'Toole KM, Sawczuk IS (1990) Carcinoma in situ of the testes: diagnosis by aspiration flow cytometry. J Urol 143:359–361
Nielsen H, Nielsen M, Skakkebæk NE (1974) The fine structure of possible carcinoma-in-situ in the seminiferous tubules in the testis of four infertile men. Acta Pathol. Microbiol. Scand [A] 82:235–248
Nistal, M, Codesal J, Paniagua R (1989) Carcinoma in situ of the testis in infertile men. A histological, immunocytochemical, and cytomorphometric study of DNA content. J Pathol 159:205–210
Nüesch-Bachmann JH, Hedinger C (1977) Atypische Spermatogonien als Präkanzerose. Schweiz. Med Wochenschr 107:795–801

Olsson L, Rahbek Sørensen H, Behnke O (1984) Intratumoral phenotypic diversity of cloned human lung tumor cell lines and consequences for analyses with monoclonal antibodies. Cancer 54:1757–1765

Oosterhuis JW, Castedo SMM, De Jong B, Cornelisse CJ, Dam A, Sleijffer DT, Schraffordt Koops, H (1989) Ploidy of primary germ cell tumors of the testis. Pathogenetic and clinical relevance. Lab Invest 60:14–21

Pedersen KV, Boiesen P, Zetterlund CG (1987) Experience of screening for carcinoma-in-situ of the testis among young men with surgically corrected maldescended testes. Int J Androl 10:181–185

Pierce GB, Abell MR (1970) Embryonal carcinoma of the testis. Pathol Annu 5: 27–32

Pottern LM, Brown LM, Hoover RN, Javadpour N, O'Connell KJ, Statzman RE, Blattner WA (1985) Testicular cancer risk among young men: role of cryptorchidism and inguinal hernia. JNG'l 74:377–381

Pryor JP, Cameron KM, Chilton CP, Ford TF, Parkinson MC, Sinokrot J, Westwood, CA (1983) Carcinoma in situ in testicular biopsies from men presenting with infertility. Br J Urol 55:780–784

Raghavan D, Sullivan AL, Peckham NJ, Munro Neville A (1982) Elevated serum alphafetoprotein and seminoma: clinical evidence for histologic continuum? Cancer 50:982–989

Rowley MJ, Heller CG (1966) The testicular biopsy: surgical procedure, fixation, and staining techniques. Fertil. Steril. 17:177–186

Scheiber K, Ackerman, D, Studer, UE (1987) Bilateral testicular germ cell tumors: a report of 20 cases. J Urol 138:73–76

Schottenfeld D, Warshauer ME, Sherlock S, Zauber AG, Leder M, Payne R (1980) The epidemiology of testicular cancer in young adults. Am J Epidemiol 112: 232–246

Schütte B (1988) Early testicular cancer in severe oligozoospeermia. In: Holstein AF, Leidenberger F, Hölzer KH, Bettendorf G (eds) Advances in andrology. Diesbach, Berlin, pp 188–190 (Carl Scerirren Symposium)

Shalet SM, Beardwell CG, Jacobs HS, Pearson D (1978) Testicular function following irradiation of the human prepubertal testis. Clin Endocrinol 9:483–490

Skakkebæk NE (1972) Possible carcinoma-in-situ of the testis. Lancet 2:516–517

Skakkebæk NE (1978) Carcinoma in situ of the testis: frequency and relationship to invasive germ cell tumours in infertile men. Histopathology 2:157–170

Skakkebæk NE, Berthelsen JG (1981) Carcinoma-in-situ of the testis and invasive growth of different types of germ cell tumours. A revised germ cell theory. Int J Androl [Suppl] 4:26–34

Skakkebæk NE, Berthelsen JG, Müller J (1982) Carcinoma-in-situ of the undescended testis. Urol Clin North Am 9:377–385

Skakkebæk NE, Berthelsen JG, Giwercman A, Müller J (1987) Carcinoma-in-situ of the testis: possible origin from gonocytes and precursor of all types of germ cell tumours except spermatocytoma. Int J Androl 19:19–28

Swerdlow AJ, Wood KH, Smith PG (1983) A case-control study of the aetiology of cryptorchidism. J Epidemiol. Community Health 37:238–244

Thompson J, Williams CJ, Whitehouse JMA, Mead GM (1988) Bilateral testicular germ cell tumours: an increasing incidence and prevention by chemotherapy. Br Med J 62:374–376

Von der Maase H, Berthelsen JG, Jacobsen GK, Hald T, Rørth M, Christophersen IS, Sørensen BL, Walbom-Jørgensen S, Skakkebæk NE (1985) Carcinoma-in-situ of testis eradicated by chemotherapy. Lancet 1:98

Von der Maase H, Rørth M, Walbom-Jørgensen S, Sørensen BL, Christophersen IS, Hald T, Jacobsen GK, Berthelsen JG, Skakkebæk NE (1986a) Carcinoma in situ of the contralateral testis in patients with testicular germ cell cancer. A study of 27 cases in 500 patients. Br Med J 293:1398–1401

Von der Maase H, Giwercman A, Skakkebæk NE (1986b) Radiation treatment of carcinoma-in-situ of testis. Lancet 1:624–625

Von der Maase H, Giwercman A, Müller J, Skakkebæk NE (1987) Management of carcinoma-in-situ of the testis. Int J Androl 10:209–220

Von der Maase H, Meinecke B, Skakkebæk NE (1988) Residual carcinoma-in-situ of contralateral testis after chemotherapy. Lancet 1:477–478

Walt H, Emmerich P, Cremer T, Hofmann M-C, Bannwart F (1989) Supernummerary chromosome 1 in interphase nuclei of atypical germ cell in paraffin embedded human seminiferous tubules. Lab Invest 61:527–531

Who (1987) Laboratory manual for the examination of human semen and semen-cervical mucus interaction Cambridge University Press, Cambridge

# *Characterization of Precancerous and Neoplastic Human Testicular Germ Cells*

H. Walt,[1] P. Emmerich,[2] A. Jauch,[2] and C.D. DeLozier-Blanchet[3]

[1] Department of Gynecology and Obstetrics, Research Division of Gynecologic Oncology, University Hospital, 8091 Zürich, Switzerland
[2] Institute of Human Genetics and Anthropology, University of Heidelberg, Im Neuenheimer Feld 328, W-6900 Heidelberg, FRG
[3] University Institute of Medical Genetics, CMU, 9 Avenue de Champel, 1211 Geneva, Switzerland

## Introduction

Testicular germ cell tumors are in most cases preceded by an intratubular noninvasive form of neoplasia, known as carcinoma in situ (CIS, Skakkebaek and Berthelsen 1978). These atypical germ cells resemble gonocytes or spermatogonia (Nüesch-Bachmann and Hedinger 1977), but also show nuclear atypia indicative of malignant transformation. Atypical intratubular cells have been known to precede overt tumors for up to 16 years (Bannwart et al. 1988). However, they do not have invasive properties and the attempts to grow them as xenografts in nude mice have been unsuccessful (Walt and Stevens unpublished results). The transition of CIS to invasive carcinoma therefore remains poorly understood. In an effort to understand better the pathobiology of testicular neoplasias and their precancerous precursors, we have studied the expression of certain biochemical tumor markers (Walt et al. 1986) and of alkaline phosphatase isozymes (Hofmann et al. 1989), the activity of nucleolus organizer regions (NORs) (Delozier-Blanchet et al. 1986a), and the presence of a number of specific chromosomal regions by in situ hybridization on paraffin sections (Emmerich et al. 1989). For these investigations we have used xenografts and their related cell lines (Delozier-Blanchet et al. 1986b; Emmerich et al. 1989; Hofmann et al. 1989; Walt et al. 1986) as well as primary tumors and primary precancerous and ejaculated germ cells (Jauch et al. 1989; Walt et al. 1989). We hope that this summary of data will allow the development of new strategies for further research in this area.

## Cytogenetic Analysis and Assessment of Nucleolus Organizer Regions in Germ Cell Neoplasia

Testicular germ cell tumors express a number of consistent structural and numerical chromosomal alterations (Delozier-Blanchet et al. 1986b; Castedo

Recent Results in Cancer Research, Vol. 123

et al. 1989). The most prominent and consistent alteration is the marker chromosome i(12p), which can be detected in human germ cell tumors of both sexes (Atkin and Baker 1983; Emmerich et al. 1989).

Analysis of NORs provides another insight into the chromosomal alterations in human germ cell neoplasia. These repetitive DNA regions, located on the short arms of the acrocentric chromosomes (groups D and G) in man, appear to be particularly subject to alterations in their number, activity, and/or location in certain malignant cells (Henderson and Megraw-Ripley 1982; Youngshan and Stanley 1988).

We are currently investigating by cytogenetic and molecular methods the incidence of ectopic or other NOR alterations in germ cell malignancies. Active NORs are detected by silver staining, and their chromosomal location is determined by simultaneous Giemsa-trypsin banding. The search for displaced NORs which may not be detected by cytogenetic methods because they are not active in the particular cell type studied involves the application of 18s and 28s rRNA probes to metaphase spreads by in situ hybridization techniques.

We found that four of seven surgical specimens or tumor cell lines derived from embryonal carcinoma of the testis contained active NORs in ectopic locations (Delozier-Blanchet et al. 1986a), as determined by silver staining techniques, whereas normal tissues from the corresponding patients did not. Two other testicular tumors studied since, as well as an ovarian dysgerminoma, had ectopic NORs in a minority of mitoses. This raises the question of a potential cause and effect relationship between their translocation and potentially modified gene expression related to malignant progression. In support of the hypothesis that NOR distribution and rearrangement may be important in tumor evolution, Takahashi et al. (1986) showed that translocation of the gene for c-*abl* to a NOR was associated with oncogene activation in a rat leukemia cell line. There is, however, no molecular explanation for the occurrence of active NORs in an ectopic location in germ cell tumors.

## Interphase Cytogenetics in the Analysis of Germ Cell Precancerous States

In situ hybridization techniques that make possible the analysis of numerical and structural aberrations in interphase as well as metaphase nuclei have been developed recently (Cremer et al. 1986, 1988). Using a modified procedure we have demonstrated that hybridized signals can be detected even in paraffin sections (Emmerich et al. 1989). This methodology is particularly useful for the assessment of numerical chromosome aberrations in paraffin sections of embedded CIS cells. Techniques for the three-dimensional evaluation of chromosomal spots in CIS nuclei are, however, time consuming and need further improvement.

**Interphase Cytogenetics and Three-Dimensional Imaging**

We presently use the following protocol for preparation. For complete staining of chromosome 12, a chromosome 12 specific DNA library (LA12NS01) from the American Type Culture Collection is used. Phage DNA is amplified in the *Escherichia coli* host LE392, and phage DNA pools extracted according to standard protocols (Maniatis et al. 1982; Lichter et al. 1988). Phage DNA is nicktranslated with biotin-11-deoxyuridinetriphosphate (dUTP) as described by Langer-Safer et al. (1982). For chromosome in situ suppression (CISS) hybridization experiments the hybridization mixture contained: 5% formamide, two time standard saline citrate (2XSSC), 10% dextrane sulfate, 1 mg/ml salmon sperm DNA, 500 μg/ml human placental DNA, and 200 μm/ml biotinylated chromosome 12 specific phage-DNA pools. A quantity of 10 μm of the mixture was added to the specimen and the slide was sealed with a coverslip (18 × 18 mm). In both cases, the specimen and probe were denatured simultaneously in a moist chamber at 76°C for 10 min. Hybridization occurred overnight at 37°C.

Slides were washed at 45°C three times for 5 min each in 50% formamide/1XSSC and then at 60°C three times for 5 min each in 0.1XSSC. Detection of hybridized DNA was carried out with avidin-fluorescein isothiocyanate (FITC) and signals were amplified with biotinylated antiavidin as described previously by Cremer et al. (1988). A further step will be double hybridization experiments with biotinylated DNA from chromosome 12 specific libraries and differently labeled cosmide clones, derived from the short arm of chromosome 12. This should lead to a characteristic staining of the marker chromosome i(12p), not only in metaphase spreads but also in interphase nuclei.

The most promising cells for a noninvasive search for aneuploid precancerous elements are expected to be semen samples. That these cells may indicate at an early stage a developing germ cell tumor, after total DNA analysis of ejaculated cells, has recently become evident (Giwercman et al. 1988).

Using a chromosomal probe specific for the centromeric region of chromosome 1, *pUC1.77*, we were able to detect numerical chromosomal aberrations in ejaculated germ cells which were similar to those found in paraffin sections (Jauch et al. 1989). The preliminary data showed supernumerary signals in nuclei from patients with and without tumors (Fig. 1). More semen samples will have to be analyzed in order to identify candidates for germ cell neoplasias among those at-risk men between 20 and 35 years of age. Initial experiences with flow cytometric analyses are still not satisfactory in this respect. More sophisticated hybridization techniques combined with confocal-laser-scanning microscopy are promising, however. Domains of chromosome 12 can be localized simultaneously and identified (Fig. 2) very accurately and quickly, as shown with leukocyte controls.

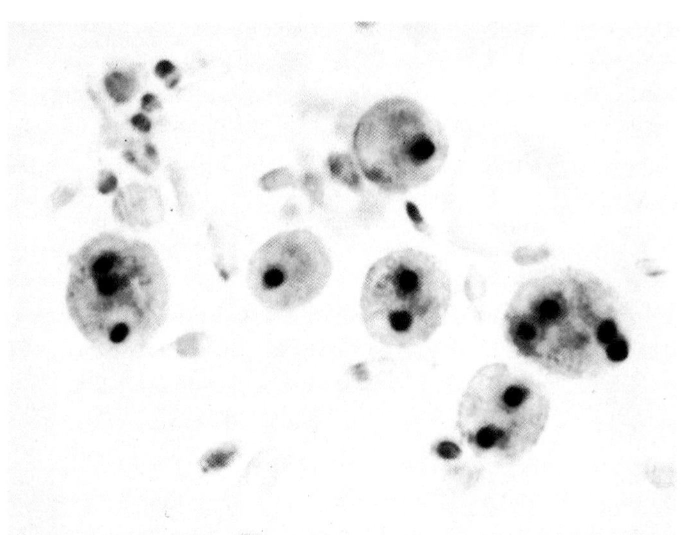

**Fig. 1.** Nuclei from exfoliated germ cells present in semen from a patient with oligoasthenoteratozoospermia (OAT) syndrome shows one, two, three, and four spots after in situ hybridization with DNA probe *pUC1.77* followed by streptavidin-peroxidase detection

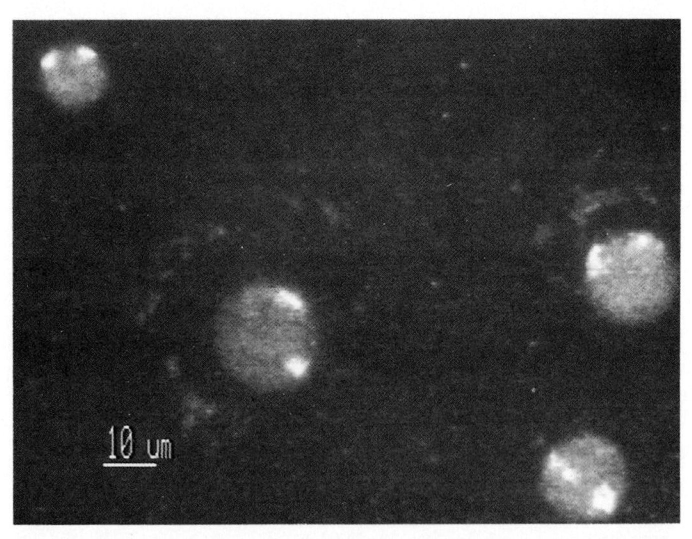


**Fig. 2.** Four interphase nuclei from a human lymphocyte culture (46, XY) after chromosome in situ suppression hybridization with biotinylated inserts of a chromosome 12 specific phage library (avidin-FITC detection). Two clearly separated chromosome domains (bright fluorescence) are visible. Nuclei were counterstained with propidium iodide and analyzed with a confocal laser scanning microscope (Leica CLSM). Pictures were stored on optical disk and recorded on a Sony UP-5000P color video printer

DNA probes specific for the i(12p) marker chromosome will make it possible to investigate the genesis of this marker directly within the tissue where it originates and most notably in CIS cells. Furthermore, this method could be a useful tool for the identification of metastases.

## Comment

There is growing evidence that some solid tumors originate from a tetraploid stage (Shackney et al. 1989) and may evolve into aneuploid neoplasia through loss of chromosome material. This may also be true for human testicular germ cell neoplasia (Oosterhuis et al. 1989). A close pathogenetic relationship between seminoma and embryonal carcinoma has been postulated, and our findings with borderline tumors having features of seminoma and embryonal carcinoma strongly support this concept (Walt et al. 1986). This tumor existed as two forms. One was firm, more seminoma like and presented more total DNA than the second, which was soft and histologically resembled embryonal carcinoma. Additional studies of similar cases could provide new insight into the relatedness of seminomas and embryonal carcinomas.

A hypothesis of early testicular tumor development is provided in Fig. 3. We propose two important steps in the formation of CIS cells; first, tetra-

| Type of germ cells / DNA characteristics | Primordial germ cells | Spermatogonia | Binucleated, "dysplastic" germ cells | Atypical germ cells | Seminoma cells | Cells of embryonal carcinoma |
|---|---|---|---|---|---|---|
| Ploidy | Euploid | Euploid | Euploid / Aneuploid | Aneuploid | Aneuploid | Aneuploid |
| Number of chromosome 1 in interphase nuclei | ? | 2 | ? | 2, 3, 4 | 2, 3, (4) | 2, 3 |
| Modal chromosome number | 46, XY | 46, XY | ? | 70-80, XY | 70-80, XY | 55-65, XY |

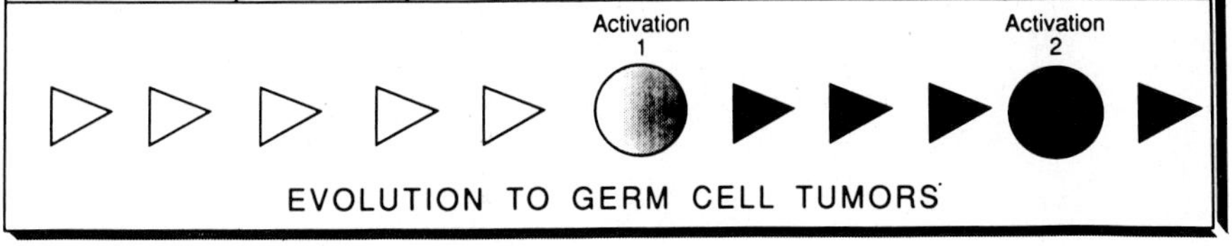

Activation 1

Activation 2

EVOLUTION TO GERM CELL TUMORS

**Fig. 3.** Hypothetical scheme of human testicular germ cell tumor development. Initial elements for tumor progression (activation 1, intratubular) are considered to be bi- or multinucleated spermatogonia, in which nuclear fusion might occur later and produce aneuploid CIS. Activation (2) is an extratubular event on the way from seminoma to embryonal carcinoma. Loss of genomic material is suggested to be a prerequisite for evolution to a more metastatic tumor form

ploidization of the genome of normal spermatogonia and, second, abnormal segregational events leading of loss of (specific?) chromosomes associated with structural chromosomal aberrations. In addition to the cytogenetically evident aberrations involved in tumorigenesis, submicroscopic genetic changes are expected to play a major role in multistage carcinogenesis. Additional cytogenetic studies are, however, needed to confirm this hypothesis.

Mechanisms which may lead to tetraploidization of spermatogonia are presently not well delineated. Possibilities include endomitotic events as well as binucleation with subsequent nuclear fusion. In the latter case an abnormal number or function of intercellular bridges between spermatogonia may contribute to cellular fusion. Our experience is that bi- and multinucleated spermatogonia often appear in company with CIS in the seminiferous tubules of patients with and without tumors as well as in cryptorchid testes (data not shown). In addition, increased numbers of binucleated cells have been found in testicular biopsies of patients with impaired spermatogenesis following chemical or X-ray treatment (Schulze 1981). It is not known whether these patients are at increased risk of developing germ cell tumors.

Detection of aneuploid cells in the semen of patients with CIS as well as with manifest germ cell tumors has recently been reported (Giwercman et al. 1988). Interphase cytogenetics of exfoliated testicular germ cells may thus provide an early noninvasive approach for the detection of CIS and/or tumor cells in patients with, for example, infertility problems, who are at increased risk of developing testicular cancer.

*Acknowledgements.* We thank Professor D. Hauri, Dept. of Urology, University Hospital, Zürich, for continuing support during numerous research projects. A part of the results were presented as a poster at the IVth International Congress of Andrology, May 14th–18th, 1989, Florence, Italy. The study was supported by the Krebsliga des Kantons Zürich and the Swiss National Science Foundation. We gratefully acknowledge the Leica Ltd. for supplying the confocal laser scanning microscope.

### *References*

Atkin NB, Baker MC (1983) i(12p): Specific chromosomal marker in seminoma and malignant teratoma of the testis? Cancer Genet Cytogenet 10:199–204

Bannwart F, Sigg C, Hedinger C (1988) Morphologie, Biologie und therapeutische Konsequenzen der atypischen Keimzellen des Hodens. Zentralbl Haut Geschlechtskr 154:861–866

Castedo SMMJ, De Jong B, Oosterhuis JW, et al. (1989) Cytogenetic analysis of ten seminomas. Cancer Res 49:439–443

Cremer T, Landegent J, Brückner A, Scholl HP, Schardin M, Hagar HD, Devilee P, Pearson P, Ploeg M (1986) Detection of chromosome aberrations in the human interphase nucleus by visualization of specific target DNAs with radioactive and nonradioactive in situ hybridization techniques: diagnosis of trisomy 18 with probe LI.84. Hum Genet 74:346–352

Cremer T, Tesin D, Hopman AHN, Manuelidis (1988) Rapid interphase and metaphase assessment of specific chromosomal changes in neuroectodermal tumor cells by in situ hybridization with chemically modified DNA probes. Exp Cell Res 176:199–220

Damjanov I (1984) Recent advances in the understanding of the pathology of the testicular germ cell tumors. World J Urol 2:12–17

Delozier-Blanchet CD, Walt H, Engel E (1986a) Ectopic nucleolus organizer regions (NORs) in human testicular tumors. Cytogenet Cell Genet 4:107–113

Delozier-Blanchet CD, Walt H, Engel E (1986b) Cytogenetic studies of human testicular germ cell tumours. Int J Androl 10:69–77

Emmerich P, Jauch A, Hofmann MC, Cremer T, Walt H (1989) Interphase cytogenetics in paraffin embedded sections from human testicular germ cell tumor xenografts and in corresponding cultured cells. Lab Invest 61:235–242

Giwercman A, Clausen OPF, Skakkebaek NE (1988) Carcinoma in situ of the testis: aneuploid cells in semen. Br Med J 296:1762–1764

Goodpasture C, Bloom SE (1975) Visualization of nucleolar organizer regions in mammalian chromosomes using silver staining. Chromosoma 53:37–50

Henderson AS, Megraw-Ripley S (1982) Rearrangement in rDNA-bearing chromosomes in cell lines from neoplastic cells. Cancer Genet Cytogenet 6:1–16

Hofmann MC, Jeltsch W, Brecher J, Walt H (1989) Alkaline phosphatase isozymes in human testicular germ cell tumors, their precancerous stage and three related cell lines. Cancer Res 49:4696–4700

Jauch A, Walt H, Emmerich P, Sigg C, Cremer T (1989) Aneuploid premeiotic cells in semen from patients with fertility problems and testis tumors detected by interphase cytogenetics. Serono Symp Rev [Suppl] 1:301

Langer-Safer PR, Levine M, Ward DC (1982) Immunological method for mapping genes on Drosophila polytene chromosomes. Proc Natl Acad Sci USA 79: 4381–4385

Lichter P, Cremer T, Borden J, Manuelidis L, Ward DC (1988) In situ delineation of individual chromosomes in normal metaphase spreads and interphase nuclei using recombinant DNA libraries. Hum Genet 80:224–234

Maniatis T, Fritsch EF, Sambrook J (1982) Molecular cloning: a laboratory manual. Cold Spring Harbor Laboratory, New York

Nüesch-Bachmann IH, Hedinger C (1977) Atypische Spermatogonien als Präkanzerose. Schweiz Med Wochenschr 107:795–801

Oosterhuis JW, Castedo SMMJ, De Jong B, Cornelisse CJ, Dam A, Sleyfer DT, Schraffordt Koops H (1989) Ploidy of primary germ cell tumors of the testis. Lab Invest 60:14–21

Schulze W (1981) Normal and abnormal spermatogonia in the human testis. Fortschr Androl 7:33–45

Shackney SE, Smith CA, Miller SBW, Burholt DR, Murtha K, Giles HR, Ketterer DM, Pollice AA (1989) Model for the genetic evolution of human solid tumors. Cancer Res 49:3344–3354

Skakkebaek NE, Berthelsen JG (1978) Carcinoma in situ and orchiectomy. Lancet 2:204–205

Takahashi R, Koichiro M, Maeda S (1986) Secondary activation of c-*abl* may be related to translocation to the nucleolar organizer region in an in-vitro-cultured rat leukemia cell line (K3d). Proc Natl Acad Sci USA 83:1079–1083

Walt H, Hedinger C (1984) Differentiation of human testicular embryonal carcinoma and teratocarcinoma grown in nude mice and soft-agar cultures. Cell Differ 15: 81–86

Walt H, Arrenbrecht S, Delozier-Blanchet CD, Keller PJ, Nauer R, Hedinger CE (1986) A human testicular germ cell tumor with borderline histology between

seminoma and embryonal carcinoma secreted beta-human chorionic gonadotropin and alpha-fetoprotein only as a xenograft. Cancer 48:139–146
Walt H, Emmerich P, Cremer T, Hofmann MC, Bannwart F (1989) Supernumerary chromosome 1 in interphase nuclei of atypical germ cells in paraffin-embedded human seminiferous tubules. Lab Invest 61:527–531
Youngshan Z, Stanley WS (1988) Effect of differentiating agents on nucleolar organizer activity in human melanoma cells. Cancer Genet Cytogenet 31:253–262

# *Cytogenetic Investigation of Gonadal Carcinomas in Situ in Syndromes of Abnormal Sexual Differentiation*

I. Vorechovsky[1] and K. Mazanec[2]

[1] Research Institute of Child Health, Černopolní 9, 662 62 Brno, Czechoslovakia
[2] 2nd Department of Pathology, Children's Teaching Hospital, Černopolní 9, 662 63 Brno, Czechoslovakia

## Introduction

The development of frequently occurring gonadal tumors in some syndromes of abnormal sexual differentiation (Verp and Simpson 1987) is often preceded by carcinoma in situ (CIS) (Skakkebaek et al. 1987). Although the finding of CIS is believed to warrant gonadectomy, the CIS pattern may persist with no clinical manifestations for a very long time. To detect early changes in chromosome number and/or structure in the gonadal tissue at risk of cancer we studied the gonads of four intersex patients.

## Patients and Methods

A prophylactic gonadectomy was performed in one patient with Swyer's syndrome, one girl with the complete form of testicular feminization syndrome (TFS), and two patients with the incomplete form of TFS at the ages of 10, 13, 15, and 18 years, respectively. Chromosomal analysis of 100 mitotic peripheral lymphocytes in each patient revealed nonmosaic karyotype 46, XY. Histopathological examinations of the gonads showed a typical CIS pattern in the patient with Swyer's syndrome and in the patient with the complete form of TFS (Mazanec and Vorechovsky 1989).

Direct cytogenetic investigations were attempted on all samples but no mitoses were obtained. The remaining parts of specimens were disaggregated mechanically, and two were treated in collagenase (ÚSOL, Czechoslovakia) solution. The cells were grown in serum-supplemented MEM (Sevac) with 100 IU penicillin and 100 μg streptomycin/ml. Only the first four passages were harvested. Colcemid (Serva) was added for the last 2 h of culture. The cells were treated with hypotonic solution (0.075 *M* KCl) for 20 min at 37°C, and fixed in three changes of methanol/acetic acid (3 : 1, v/v). The cell suspension was spread on cold slides and air dried. Chromosomes were

Recent Results in Cancer Research, Vol. 123

stained conventionally with Giemsa, and some preparations were G-banded with trypsin (Spofa). For each patient, 33–109 mitotic cells were analyzed.

## Results and Discussion

In all patients a bimodal distribution of chromosome numbers with peaks at diploid and near-tetraploid values was found. Examination of 102 cells derived from the patient with Swyer's syndrome showed 22 near-tetraploid mitoses including 6 endoreduplicated cells, 5 of them incomplete (Fig. 1). The missing chromosomes could not be identified exactly. A possible specific cytogenetic marker of the gonadal germ cell tumors, isochromosome 12p (Atkin and Baker 1983; Delozier-Blanchet et al. 1987), was not detected in any cell. Chromosome Y was identified in one copy of all the mitoses observed. Premature chromosome condensation was seen in 1 of 94 mitoses studied in the patient with Swyer's syndrome.

The increased proportion of near-tetraploid cells was similar to that found in other CIS, e.g., cervix uteri (Grandberg 1971). Although some environmental influences were shown to induce tetraploidy in culture, several studies showing a relatively stable percentage of near-tetraploid cells in consecutive passages indicate that a potential to form tetraploid cells is genetically determined and the increased number of tetraploid cells is considered a putative biomarker for genetic predisposition to several types of cancer (Danes et al. 1986). The findings of heteroploid cells with the diplochromosome figures, which are typical of endoreduplications (Fig. 1), seem to support the view that tetraploid cells may represent unstable systems with frequent chromosome loss leading to the near-triploid subpopulations frequently encountered in the testicular tumors (see Fig. 2). The major mechanism leading to tetraploidy appears to be an endoreduplication, but others are possible, cell fusions in vivo being among those which have been much discussed (Kovacs 1985; Munzarova and Kovarik 1987). The described developmental mechanism (Fig. 2) need not be confined to the gonadal tumors, but it may be more common, at least in some solid tumors. The polyploidization-segregation process leading to the loss of heterozygozity was also supported by the karyotypic and southern hybridization analyses of retinoblastoma susceptibility gene in the development of osteosarcoma (Toguchida et al. 1988), rather than homologous recombination or nondisjunction-duplication, which is known to be the most frequent mechanism in diploid tumors such as retinoblastoma or Wilms' tumor. It is also possible to support the view that the potential of tissue to form an increased number of tetraploid cells in vitro may be an expression of genetic instability (Hsu 1983), which may be associated with a higher probability of mutations and enhanced risk of malignant transformation.

Direct cytogenetic examination of testicular tissue in one patient with oligospermia and CIS of the right testicle revealed nine mitoses with similar

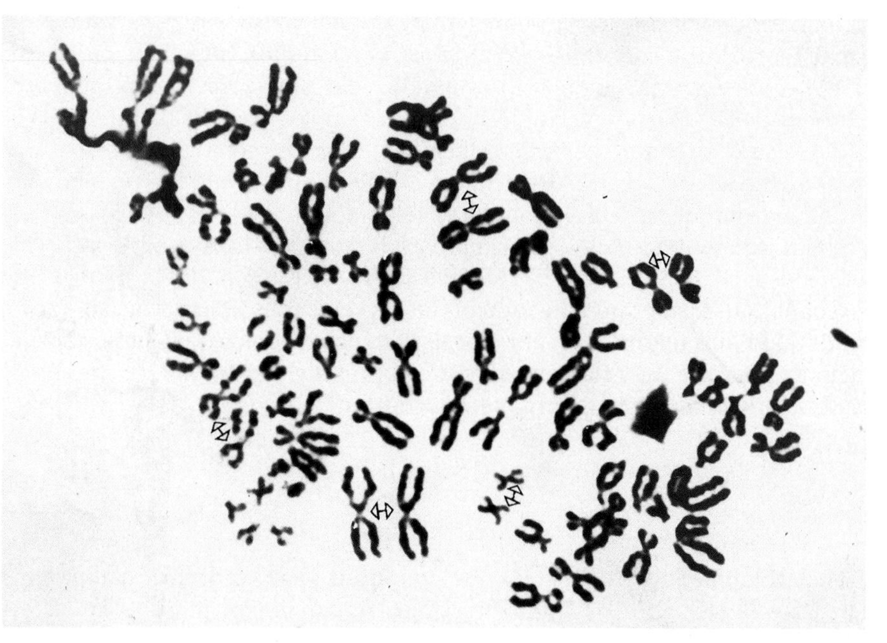

**Fig. 1.** Hypotetraploid mitotic cell with the diplochromosome patterns (◁—▷) typical of endoreduplication. Conventionally stained

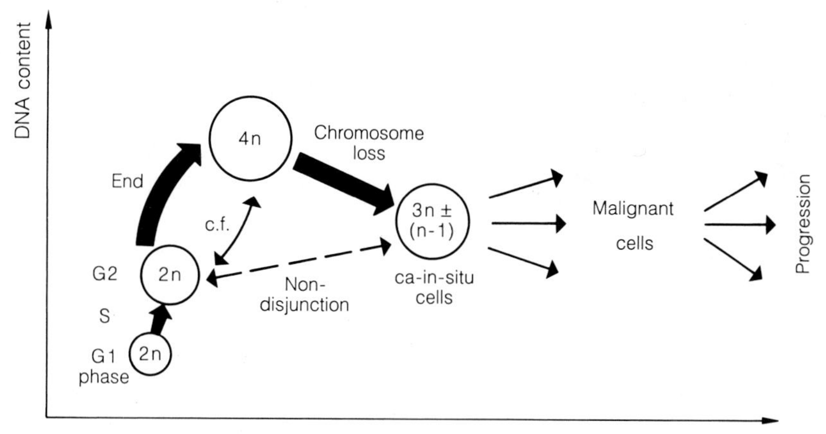


**Fig. 2.** Possible mechanism contributing to the development of some solid tumors. *end*, endoreduplication/endomitosis; *c.f.*, cell fusion; *n*, haploid set

broad variations of chromosome numbers and consistent observations of small metacentric and submetacentric marker chromosomes (Lehmann et al. 1986). However, the metacentric markers did not seem to represent isochromosomes. The absence of i(12p) in our samples suggests that the marker is associated with more advanced clinical stages of the gonadal tumors, as was proposed earlier (Delozier-Blanchet et al. 1987), or it could be confined only to a subgroup of testicular tumors (Castedo et al. 1988).

Although our methodological approach is much less likely to detect specific chromosomal changes in the testicular CIS cells than direct cytogenetic examination, further investigation of this precancerous tissue, which is known to develop into tumors with more or less specific chromosomal marker, could help to elucidate the relationship between primary and secondary alterations and their role in the long-term carcinogenesis.

## Conclusion

The development of frequent gonadal tumors in some syndromes of abnormal sexual differentiation is often preceded by carcinoma in situ. To detect possible early changes in chromosome number and/or structure associated with the carcinoma in situ, gonadal tissue was studied in one patient with the complete form of TFS, two patients with the incomplete form of this syndrome, and one patient with Swyer's syndrome. In all patients, aneuploidy with a bimodal distribution of chromosome numbers with the peaks at diploid and near-tetraploid values was shown. Possible specific chromosomal marker of gonadal tumors, i(12p), was not found. Its absence suggests that this marker might be associated with more advanced stages of long-term cancer development or it could be confined only to a subgroup of testicular tumors. The hypothesis that the major source of hypotetraploid or near-triploid cells often found in many gonadal tumors might be endoreduplications with their subsequent chromosome loss is supported.

## *References*

Atkin NB, Baker MC (1983) i(12p): specific chromosomal marker in seminoma and malignant teratoma of the testis? Cancer Genet Cytogenet 10:199–204

Castedo SMMJ, deJong B, Oosterhuis JW (1988) i(12p) – negative testicular germ cell tumours: a different group? Cancer Genet Cytogenet 35:171–178

Danes BS, Boyle PD, Traganos F, Melamed MR (1986) A standardized assay to identify colon cancer genotypes by in vitro tetraploidy in human dermal fibroblasts. Dis Markers 4:271–282

Delozier-Blanchet CD, Walt H, Engel E, Vuagnat P (1987) Cytogenetic studies of human testicular germ cell tumours. Int K Androl 10:69–78

Grandberg I (1971) Chromosomes in preinvasive, microinvasive and invasive cervical carcinoma. Hereditas 68:165

Hsu TC (1983) Genetic instability in the human population: a working hypothesis. Hereditas 98:1

Kovacs G (1985) Premature chromosome condensation: evidence for in vivo cell fusion in human malignant tumours. Int J Cancer 36:637–641

Lehmann D, Temminck B, Litman K, Leibundgut B, Hadziselimovic F, Muller H (1986) Autoimmune phenomena and cytogenetic findings in a patient with carcinoma (seminoma) in situ. Cancer 58:2013–2017

Mazanec K, Vorechovsky I (1989) Gonadal histopathology in some syndromes of abnormal sexual differentiation. Cesk Patol 25:54

Munzarova M, Kovarik J (1987) Is cancer a macrophage-mediated autoaggressive disease? Lancet 1:952–954

Skakkebaek NE, Berthelsen JG, Giwercman A, Muller J (1987) Carcinoma-in-situ of the testis: possible origin from gonocytes and precursors of all types of germ cell tumours except spermatocytoma. Int J Androl 10:19–28

Toguchida J, Ishizaki K, Sasaki MS, Ikenaga M, Sugimoto M, Kotoura Y, Yamamuro T (1988) Chromosomal reorganization for the expression of recessive mutation of retinoblastoma susceptibility gene in the development of osteosarcoma. Cancer Res 48:3939–3943

Verp MS, Simpson JL (1987) Abnormal sexual differentiation and neoplasia. Cancer Genet Cytogenet 25:191–218

# Cell Lines

## *Human Embryonal Carcinoma and Yolk Sac Carcinoma in Vitro: Cell Lineage Relationships and Possible Paracrine Growth Regulatory Interactions*

M.F. Pera, S. Cooper, W. Bennet, and I. Crawford-Bryce

Department of Zoology, Cancer Research Campaign, University of Oxford, Oxford, OX1 3PS, Great Britain

### Introduction

In this chapter, we will review some aspects of work in our laboratory on the control of growth and differentiation in human teratocarcinoma stem cells. We begin by describing the three cell types which we have isolated in vitro from human teratomas: embryonal carcinoma, yolk sac carcinoma resembling visceral endoderm (solid yolk sac carcinoma), and yolk sac carcinoma resembling parietal endoderm (endodermal sinus tumour). Next, we discuss the development of a new panel of monoclonal antibodies for the study of cell differentiation lineage in human teratomas. We then describe properties of a spontaneously differentiating multipotent clone of human embryonal carcinoma, and we present evidence of possible paracrine growth interaction between teratocarcinoma stem cells and yolk sac carcinoma cells.

### Cultured Cell Lines from Human Teratomas

Andrews et al. (1980, 1983) provided the first definitive characterisation of human embryonal carcinoma stem cells grown in vitro, based upon the properties of several long-established cell lines. Detailed studies of clonal derivatives of the cell line Tera 2 further clarified the properties of human embryonal carcinoma (Andrews et al. 1984 Thompson et al. 1984), and the phenotype of the cultured cells generally conformed to results obtained directly on biopsy specimens (Andrews and Damjanov 1985).

Our work with a new panel of cell lines confirmed and extended these earlier observations. Using mesenchymal feeder cell support, we established a panel of cell lines from human testicular teratomas. About two-thirds of biopsies gave rise to permanent cell lines in our hands (Pera et al. 1987). The cell type established in culture most frequently from biopsy specimens showed

Recent Results in Cancer Research, Vol. 123

properties consistent with those of human embryonal carcinoma. These properties included cell surface expression of stage-specific embryonic antigens (SSEAs)-3 and -4, but not -1; low levels of class I major histocompatibility complex antigen expression; expression of cytokeratins 8, 18 and 19 characteristic of simple epithelia; staining at the cell borders with antibodies against desmosomal plaque proteins; low levels of expression of extracellular matrix molecules fibronectin, laminin and collagen type IV; and occasional secretion of modest amounts of α-fetoprotein and human chorionic gondatropin into the culture medium. When injected into nude mice, these cell lines gave rise to tumours composed chiefly of undifferentiated embryonal carcinoma. Cells expressing human chorionic gonadotropin or α-fetoprotein were seen in xenografts from several of the cell lines, and one cell line (GCT 27) showed extensive somatic differentiation, in addition to producing these markers of placental and yolk sac differentiation.

The phenotypes of our embryonal carcinoma cell lines generally resembled those of multipotent clones derived from the Tera 2 cell line, but differed in many respects from pluripotent cells of the preimplantation embryo or primordial germ cells. As yet, the cell type in normal embryos which corresponds to the human embryonal carcinoma stem cell remains unknown. We have argued that there is a close similarity between the phenotype of human embryonal carcinoma and those of tumour cells representative of extra-embryonic endodermal lineages (yolk sac carcinoma). Yolk sac differentiation is certainly prominent in the human tumours, so it is possible that human embryonal carcinoma represents a yolk sac progenitor cell which often retains multipotentiality. But until an analogous cell type in the mammalian embryo is identified, it remains difficult to discuss human embryonal carcinoma in relation to early development.

A second cell type, present in perhaps 10%–25% of teratoma biopsies, resembles in some ways rodent visceral endoderm, and is typified by cell line GCT 72. In culture, GCT 72 cells are large and flat with a vacuolated cytoplasm and distinct cell borders. At the cell surface, GCT 72 cells expressed SSEAs-3, -4 and -1; expression of class I histocompatibility was low or absent; the cytoskeleton contained cytokeratins 8, 18 and 19; desmoplakins were strongly stained at the cell borders; and abundant fibronectin, laminin and type IV collagen were secreted into the culture medium. But the most distinctive feature of this cell type is the secretion of very high levels of α-fetoprotein into the culture supernatant. Transcripts for α-fetoprotein and other markers of visceral endoderm (albumin, transferrin) were detected in total mRNA from this cell line (Krumlauf and Pera, unpublished). When injected into nude mice, GCT 72 cells formed tumours consisting of solid sheets of yolk sac carcinoma (a pattern which can be mistaken for seminoma). The solid yolk sac carcinoma histological findings are very different from the appearance of endodermal sinus tumour cells (see below). In the visceral endoderm yolk sac xenograft tumours, all cells were stained by antisera to α-fetoprotein.

The third cell type, isolated from about 20% of teratoma specimens, is a yolk sac carcinoma cell with some features of rodent parietal endoderm. These cells were mesenchymal in appearance, and were distinguished by their lack of staining with antibodies to SSEA-3 and SSEA-4, their strong coexpression (in all cultured cells) of cytokeratins and vimentin intermediate filaments, the lack of staining at cell borders with antibodies to desmoplakins, their staining by antibodies against a 180-kDa protein expressed only in cells of mesenchymal origin (Isacke and Pera, unpublished) and their secretion of the cell substrate adhesion molecules fibronectin, laminin, type IV collagen and vitronectin (see below). Most significantly, while embyronal carcinoma and visceral endoderm-type yolk sac carcinoma are strongly dependent upon serum for attachment and growth in vitro, parietal endoderm-type yolk sac carcinoma cells could attach, grow and undergo serial cultivation in vitro in the absence of serum. This type of yolk sac carcinoma cell has many features reminiscent of cells undergoing epithelial to mesenchymal transitions (Boyer et al. 1989), a phenotypic change in which epithelial cells acquire certain features of mesenchymal cells. Such transitions are associated with the appearance in epithelial cells of an individual and migratory phenotype, which is also characteristic of normal parietal endoderm (Hogan et al. 1983). When injected into nude mice, parietal endoderm-type yolk sac carcinoma cells rapidly gave rise to tumours with the histological characteristics of endodermal sinus tumour. A proportion of the xenograft tumour cells were stained with antisera to α-fetoprotein, but most were not. Endodermal sinus tumours have been associated with a poor prognosis and rapid spread in clinical studies.

## Monoclonal Antibodies for the Study of Cell Differentiation Lineage in Human Teratomas

The above data on the phenotypes of embryonal carcinoma, visceral endoderm-like yolk sac carcinoma and parietal endoderm-like yolk sac carcinoma suggested that these were a related family of epithelial cell types. The number of markers we studied was limited, however, and many were expressed in a range of cell types. To better define cell lineage relationships in the tumours, and to develop tools for the study of gene expression in stem cells, we developed a new panel of monoclonal antibodies by immunising mice with a detergent-insoluble extract of embryonal carcinoma cell line GCT 27.

From two fusions, three monoclonal antibodies were isolated which proved to be relatively limited in distribution (Pera et al. 1988). Monoclonal antibody GCTM-2 reacts with a keratan sulphate proteoglycan present in the pericellular matrix of human embryonal carcinoma cells, including clonal derivatives of Tera 2, and visceral endoderm yolk sac carcinoma cells. The purification and characterisation of this antigen have been achieved recently

(Cooper et al., submitted for publication). The intact proteoglycan has a molecular mass of about 200 kDa and was localized in the pericellular matrix of GCT 27 cells by immunoelectron microscopy. The antigen was purified from guanidine hydrochloride extracts of GCT 27 cell monolayers by anion exchange and gel filtration chromatography. When the purified antigen was digested with keratanase, or deglycosylated by chemical hydrolysis, the remaining protein core showed a molecular weight of 55 kDa. Inorganic [$^{35}$S]sulphate was incorporated into the GCTM-2 antigen during metabolic labelling of GCT 27 cells, and the incorporated label was released in a low molecular weight form by keratanase digestion.

Tissue distribution studies (Mason and Pera, this volume) showed that GCTM-2 reacted strongly with embryonal carcinoma, a proportion of seminoma, embryonic muscle, gut and placenta. Certain other epithelia stained weakly. Cornea and cartilage did not react with this antibody. These data, and the biochemical analysis of the protein, indicate that it is unlike known keratan sulphate proteoglycans.

Monoclonal antibody GCTM-3 decorated a cytoskeletal antigen present in foci of cultured GCT 27 embyronal carcinoma cells. Cloned derivatives of Tera 2 are not stained by GCTM-3. Strongest expression of the GCTM-3 cytoskeletal antigen was seen in both variants of yolk sac carcinoma. Thus this antigen is probably a marker for extraembryonic endoderm; GCT 27 has the capacity for this type of differentiation, but Tera 2 clones do not. The GCTM-3 epitope lies on a 57-kDa protein, which although it is unlikely to represent any known cytoskeletal component of human embryonal carcinoma cells, has not been further characterized as yet.

Finally, the antibody GCTM-4 reacted with an epitope which is unusual in several ways. First, this antigen was limited to expression to embryonal carcinoma cells in the panel of cells first studied. Yolk sac carcinoma cells were not stained by the antibody. Second, immunoelectron microscopy and colabelling studies localized the epitope to a fraction of the lysosomal/endosomal compartment of human embryonal carcinoma cells. The antibody reacted with a band of 69 kDa in whole cell lysates of GCT 27 cells. The nature of this antigen awaits further characterisation, but the antibody provides a relatively specific marker for EC cells.

Thus, studies on these antigens of relatively limited distribution provided further evidence of a close cell lineage relationship between human embryonal carcinoma and yolk sac carcinoma cells, since there was some overlap in expression of these markers in the three cell types. However, the antibody panel allowed clear distinction between these cell types.

**Isolation and Characterisation of a Spontaneously Differentiating, Multipotent Clone of Human Embryonal Carcinoma**

Several groups described the isolation of clonal cell lines from the human teratoma line Tera 2 which differentiate into neurons when treated with retinoic acid (Andrews et al., 1989; Thompson et al. 1984). The extensive spontaneous differentiation seen in human teratocarcinoma in patients, and in some mouse teratocarcinomas cultured in vitro, has not been observed in cultured human cell lines.

As noted above, early passage cultures of embryonal carcinoma cell line GCT 27 gave rise to teratocarcinomas in nude mice which contained a range of somatic cell types. From these xenografts, clones were isolated, and selected for the presence of multiple cell morphologies in growing colonies. These colonies were grown, cloned by single cell manipulation under microscopic guidance, and reselected for mixed morphology. One clone representative of the spontaneously differentiating variants, GCT 27 X-1, was chosen for further study and comparison with clone GCT 27 C-4, which grew in vitro as a monomorphic line consisting of only embryonal carcinoma.

Following cloning, GCT 27 X-1 cells spontaneously gave rise to morphologically altered cell types with limited growth potential. The phenotype of the stem cells in GCT 27 X-1 cultures was identical to that of the other embryonal carcinoma cell lines described previously, and to that of the GCT 27 C-4 nullipotent clones, which did not spontaneously differentiate. Continuous growth of GCT 27 X-1 stem cells was highly dependent upon mesenchymal feeder cell support, whereas nullipotent GCT 27 C-4 could grow at clonal density in the absence of a feeder layer. Both GCT 27 X-1 and GCT C-4 were aneuploid and both carried copies of isochromosome 12 p, a characteristic abnormality of germ cell tumours. Evidence for pluripotentiality of GCT 27 X-1 cells was obtained through studies of spontaneous differentiation using double-label indirect immunofluorescence. Stem cells were identified with the GCTM-2 antibody against the embryonal carcinoma proteoglycan, and expression of a range of cell surface, cytostructural and secreted markers was studied with second antibodies. In the earlier phases of differentiation, new cell types appeared with many features of extraembryonic endoderm. These cells lost the stem cell markers identified by monoclonal antibodies, and showed instead coexpression of cytokeratin and vimentin intermediate filaments, deposition of extensive quantities of extracellular matrix molecules, and expression of the 180-kDa marker of mesodermal cells. Later, as differentiation proceeded, multilayered aggregates formed in the cultures. The cells in these aggregates sometimes expressed markers characteristic of neuronal differentiation, including the neural cell adhesion molecule, and neurofilament polypeptides.

The most convincing evidence for the multipotentiality of GCT 27 X-1 cells came from xenograft studies. When injected into nude mice, GCT 27 X-1 cells gave rise to tumours consisting of embryonal carcinoma, cells expressing α-fetoprotein, endodermal sinus variant of yolk sac carcinoma, multinucleated giant cells producing large amounts of human chorionic gonadotropin, glandular epithelium, primitive mesenchyme, cartilage, squamous cysts and neuroectodermal cells. Thus GCT 27 X-1 cells were capable of differentiation into extraembryonic, as well as mesodermal, ectodermal and endodermal derivatives.

## Paracrine Growth Interactions Between Yolk Sac Carcinoma and Multipotent Embryonal Carcinoma Cells

The yolk sac is thought to perform a supporting role for the embryo proper during early development; visceral endoderm in particular secretes many of the nutritional macromolecules produced by the liver in the adult. As noted above, there is a striking contrast between the growth requirements in vitro of multipotent embryonal carcinoma cells and yolk sac carcinoma cells; the former require serum and feeder cell support, the latter require neither. We sought to determine whether or not yolk sac carcinomas produced factors which would support the growth of multipotent cells.

Parietal endoderm yolk sac carcinoma cells do not require serum for attachment and spreading in vitro. Like parietal endoderm, visceral endoderm yolk sac carcinoma cells secrete fibronectin, laminin and type IV collagen into culture medium in vitro. However, visceral yolk sac carcinoma still requires serum for attachment. Attachment of parietal endoderm type yolk sac carcinoma cells in vitro in the absence of serum is related to their secretion of vitronectin (also known as serum-spreading factor, S protein) (Cooper and Pera 1988). Vitronectin (along with fibronectin) constitutes the major cell attachment activity in serum and is specifically required for attachment and growth of human embryonal carcinoma cells in vitro. Unlike fibronectin, vitronectin is only produced by a few tissues: liver, placenta and yolk sac carcinoma. It is possible that the role of vitronectin in the early embryo may be to support migration and growth of certain cell types, including multipotent cells. A monoclonal antibody raised against umbilical cord blood vitronectin (VIT-1) immunoprecipitated a 69-kDa band from metabolically labelled parietal yolk sac carcinoma cell supernatants. The VIT-1 antibody inhibited attachment of these cells in serum-free medium. Biopsy sections of yolk sac carcinoma, liver and placenta, but no other normal tissues of mid-trimester fetuses, are stained by VIT-1. This antibody may recognise a fetal isoform of vitronectin, or a closely related protein, involved in cell adhesion.

Multipotent human embryonal carcinoma cell line GCT 27 X-1 required feeder cells for continuous growth in vitro. A factor produced by parietal endoderm yolk sac carcinoma cells replaced this feeder cell requirement. Insulin, platelet-derived growth factor, transforming growth factor β, epidermal growth factor, interleukins 1, 3 or 6, GM-CSF, bombesin, vitronectin, fibronectin and laminin all failed to replace the feeder cell requirement for GCT 27 X-1 growth. The feeder cell requirement of multipotent mouse embryonic stem cells may be replaced by the polypeptide regulator known as differentiation-inhibitory activity of leukemia inhibitory factor (Smith et al. 1988; Williams et al. 1988). However, this activity was without effect on multipotent human EC cells.

The yolk sac factor has been partially purified from serum-free supernatants of cell line GCT 44. The factor is an acidic glycoprotein with an apparent $M_r$ of approximately 70 kDa. We are continuing efforts to purify and characterise this factor in our laboratory.

These data provide support for the notion that yolk sac carcinoma cells, like their normal counterparts, secrete growth and attachment factors which function in a paracrine fashion to support multipotent stem cells. It may be that this paracrine cell interaction has a counterpart in normal development of the peri-implantation embryo.

## Conclusion

Cell lines representative of human embryonal carcinoma and two variants of yolk sac carcinoma have been isolated and characterised in vitro. Embryonal carcinoma and yolk sac carcinomas would appear to be a related family of epithelial cell types which display progressive levels of commitment to extraembryonic endodermal differentiation. A multipotent teratocarcinoma stem cell was isolated which showed the capacity for differentiation into many somatic and extraembryonic cell types. These cells resembled nullipotent cells in most respects, except that they differentiate spontaneously and showed more stringent requirements for continuous growth in vitro. Yolk sac carcinoma cells secrete vitronectin, which is required for attachment of human embryonal carcinoma cells in vitro. Yolk sac carcinoma cells also produce a factor which replaces the feeder cell requirement for growth of multipotent human teratocarcinoma stem cells. The production of growth and attachment factors by yolk sac carcinoma cells may reflect a paracrine secretory mechanism for the support of multipotent cells in the normal embryo.

*Acknowledgements.* We thank Mrs. Jean Hepburn for help in the preparation of the manuscript. Work in our laboratory is supported by the Cancer Research Campaign.

## *References*

Andrews PW, Damjanov I (1985) Immunochemistry of human teratocarcinoma stem cells. In: Sell S, Reisfeld RA (eds) Monoclonal antibodies in cancer. Humana, Clifton, pp 339–364

Andrews PW, Bronson DL, Benham F, Strickland S, Knowles BB (1980) A comparative study of eight cell lines derived from human testicular teratocarcinoma. Int J Cancer 26:269–280

Andrews PW, Goodfellow P, Damjanov I (1983) Human teratocarcinoma cells in culture. Cancer Surv 2:41–73

Andrews PW, Damjanov I, Simon D, Banting GS, Carlin C, Dracopoli NC, Fogh J (1984) Pluripotent embryonal carcinoma clones derived from the human teratocarcinoma cell line Tera-2. Lab Invest 50:147–162

Boyer B, Tucker GC, Valles AM, Franke WW, Thierry JP (1989) Rearrangements of desmosomal and cytoskeletal proteins during the transition from epithelial to fibroblastoid organization in cultured rat bladder carcinoma cells. J Cell Biol 109:1495–1509

Cooper S, Pera MF (1988) Vitronectin production by human yolk sac carcinoma cells resembling parietal endoderm. Development 104:565–574

Hogan BLM, Barlow DP, Tilly R (1983) F9 teratocarcinoma cells as a model for the differentiation of parietal and visceral endoderm in the mouse embryo. Cancer Surv 2:115–140

Pera MF, Blasco-Lafita MJ, Mills J (1987) Cultured stem cells from human testicular teratomas: the nature of human embryonal carcinoma, and its composition with two types of yolk sac carcinoma. Int J Cancer 40:334–343

Pera MF, Blasco-Lafita MJ, Cooper S, Mason M, Mills J, Monaghan P (1988) Analysis of cell differentiation lineage in human teratomas using new monoclonal antibodies to cytostructural antigens of embryonal carcinoma cells. Differentiation 39:139–149

Pera MF, Cooper S, Mills J, Parrington JM (1989) Isolation and characterization of a multipotent clone of human embryonal carcinoma cells. Differentiation 42:10–23

Smith AG, Heath JK, Donaldson DD, Wong GG, Moreau J, Stahl M, Rogers D (1988) Inhibition of pluripotential embryonic stem cell differentiation by purified polypeptides. Nature 336:688–690

Thompson S, Stern PL, Webb M, Walsh FS, Engstrom W, Evans EP, Shi WK, Hopkins B, Graham CF (1984) Cloned human teratoma cells differentiate into neuron-like cells and other cell types in retinoic acid. J Cell Sci 72:37–64

Williams RL, Hilton DJ, Pease S, Wilson TA, Stewart CL, Gearing DP, Wagner EF, Metcalf D, Nicola NA, Gough NM (1988) Myeloid leukemia inhibitory factor maintains the developmental potential of embryonic stem cells. Nature 336:684–687

# *Biochemical Analysis and Cellular Location of the GCTM-2 Antigen in Embryonal Carcinoma and in Other Tumour Cell Lines*

M.D. Mason and M.F. Pera

Institute of Cancer Research, Royal Cancer Hospital, 15 Cotswold Road, Sutton, Belmont, Surrey, SM2 5NG, Great Britain

## Cellular Location of GCTM-2 Antigen

We have previously described the development of monoclonal antibody GCTM-2 from a detergent-insoluble extract of the human embryonal carcinoma (EC) cell line GCT 27 (Pera et al. 1988). Immunofluorescence studies showed that GCTM-2 strongly stained human EC cells in preparations of cells fixed in a 50 : 50 mixture of methanol and ethanol, and in preparations utilising live cells. This indicated that it was not necessary to permeabilise the cells by fixation in order to render the antigen accessible to the antibody. Electron microscopy studies localised the antigen outside the cell surface, and these along with the above data and the resistance of the antigen to extraction with high salt and non-ionic detergent from a monolayer culture of EC were compatible with the hypothesis that the antigen was located in the pericellular matrix. It was also observed that the antigen disappeared from the cell surface during the spontaneous differentiation of human EC cells in vitro (Pera et al. 1989).

## Biochemical Analysis

Immunoblotting with GCTM-2 on whole cell lysates of GCT 27 revealed a 200-kDa band. However, when the cell monolayer was pretreated with the enzyme keratanase prior to immunoblotting, the band was shown to be degraded, resulting in a series of bands from 200 kDa down to a minimum of 55 kDa. This suggested that (a) the GCTM-2 antigen was a keratan sulphate proteoglycan, (b) that the core protein had a molecular weight close to 55 kDa and (c) that the antibody recognised an epitope on the core protein. Recent further biochemical and electron microscopic studies have confirmed the proteoglycan nature of this antigen (Cooper et al. submitted for publication).

In contrast to most of the human tumour cell lines held at the Institute of Cancer Research, which on screening against GCTM-2 by immunofluorescence did not react, we have found three human tumour cell lines that do stain using fixed cell preparations. The first, HX 18, was established from a xenograft of a human colorectal carcinoma, the second, HX 170, from a xenograft of human embryonal rhabdomyosarcoma, and the third was the well-established human choriocarcinoma cell line BeWo.

**Presence in Non-embryonal Carcinoma Cells**

Fixed cell preparations from each of these cell lines stained strongly with immunofluorescence using GCTM-2. However, live cell preparations from HX 170 and BeWo did not, indicating that in these cells permeabilistation by fixation is required to render the antigen accessible, suggesting that it may be internal. By contrast, in live cell preparations of HX 18, approximately 30% of the cell population did stain with GCTM-2, while the remaining 70% did not, indicating that in a proportion of these cells the antigen was in a similar location to that in EC.

The staining of tumour cells of gut, muscle and trophoblastic origin by GCTM-2 was not unexpected, given the apparent tissue distribution of the GCTM-2 antigen as revealed by immunohistochemistry on normal human mid-trimester fetal tissue. This is shown in Table 1.

Immunoblotting was performed on whole cell lysates of each of the three cell lines studied by immunofluorescence, both with and without prior treat-

**Table 1.** Immunohistochemistry on mid-trimester fetal tissue with GCTM-2

| | |
|---|---|
| Strongly positive | Gut epithelium |
| | Muscle (all types) |
| Moderate | Trophoblast |
| | Liver |
| | Renal distal tubules |
| Weak | Lung epithelium |
| | Bladder epithelium |
| | Oesophageal epithelium |
| | Tracheal epithelium |
| | Thyroid |
| Negative | Brain |
| | Cartilage |
| | Eye |
| | Skin |
| | Mesenchyme |
| | Testis |
| | Spleen |
| | Pancreas |
| | Thymus |
| | Lymph node |
| | Renal glomeruli |

ment of the cells with keratanase. However, unlike the results obtained with GCT 27, a band approximately 60–80 kDa in size was revealed, which was not degraded by treatment with keratanase in any of the three cell lines. As the molecular weights are close, it is possible that in these cells the antibody is detecting the core protein of the EC proteoglycan, or modifications thereof, a situation recognised in other proteoglycans (Hassell et al. 1986), although it is difficult to be certain that it is not recognising an epitope on an unrelated protein.

## Conclusions

A keratan sulphate proteoglycan has been described on differentiation in vitro of the murine F9 EC cell Line (Kapoor and Prehm 1983). Its function has not been elucidated, and it is possible that it is related to the GCTM-2 antigen in human EC. Proteoglycans have diverse functions which include binding to cell attachment factors and binding to growth factors (Ruoslahti 1988) and we are investigating these as possible functions of the GCTM-2 antigen in human EC. Finally, as new probes become available it will be possible to study in more detail the nature of the antigen in these other cell lines.

## *References*

Hassell JR, Kimura JH, Hascall VC (1986) Proteoglycan core protein families. Annu Rev Biochem 55:539–567

Kapoor R, Prehm P (1983) Changes in proteoglycan composition of F9 teratocarcinoma cells upon differentiation. Eur J Biochem 137:589–595

Pera MF, Blasco-Lafita MJ, Cooper S, Mason M, Mills J, Monaghan P (1988) Analysis of cell differentiation lineage in human teratomas using new monoclonal antibodies to cytostructural antigens of embryonal carcinoma cells. Differentiation 39:139–149

Pera MF, Cooper S, Mills J, Parrington M (1989) Isolation and Characterization of a multipotent clone of human embryonal carcinoma cells, Differentiation 42:10–23

Ruoslahti E (1988) Structure and biology of proteoglycans. Annu Rev Cell Biol 4:229–255

# *The Surface Antigen Phenotype of Human Embryonal Carcinoma Cells: Modulation Upon Differentiation and Viral Infection*

P.W. Andrews,[1] J. Marrink,[2] G. Hirka,[1] A. von Keitz,[1]
D. Th. Sleijfer,[2] and E. Gönczöl[1]

[1] Wistar Institute of Anatomy and Biology, 3601 Spruce Street, Philadelphia, PA 19104, USA
[2] Department Internal Medicine, Immunochemical Laboratory and Medical Oncology, University Hospital, 9700 RB Groningen, The Netherlands

## Introduction

The cell surface provides a rich source of developmentally controlled molecules that are of interest to several branches of basic cell biology and clinical medicine. Originating from studies of ontogeny and function in the immune system, the notion of cell surface antigen phenotype, defined by the pattern of expression of several "differentiation antigens" (Bennett et al. 1972; Boyse and Old 1978), also provides a valuable tool for following the progression of cell types in other complex differentiating systems such as those presented by teratocarcinomas. We have used the expression of a variety of surface antigens to provide objective criteria for the identification of human embryonal carcinoma (EC) cells, the stem cells of teratocarcinomas; we have also used changes in the expression of such antigens to study the differentiation of a pluripotent human EC cell line, TERA 2 (Andrews et al. 1984b). Extension of these observations to a clinical setting could provide the basis for improved histological classification of human germ cell tumors. Moreover, as we report here, the shedding of certain surface antigens characteristic of EC cells and their detection in the serum of patients could provide a useful additional means of monitoring the progress of therapy.

However, more significant, perhaps, than the value of surface antigens as operational markers of cell type is the recognition that they most likely include many molecules that function in the regulation of cell growth and differentiation – either as receptors for growth factors or as mediators of cell-cell and cell-substrate interactions. Investigation of the function and expression of developmentally regulated cell surface antigens may thus yield clues to the control of cell differentiation, not only during embryogenesis but also during oncogenesis. For example, the modulation of several glycolipid

Recent Results in Cancer Research, Vol. 123

antigens due to growth of a teratocarcinoma cell line as a xenograft in nude mice raises the question of whether the expression of these antigens, which is subject to genetic polymorphism in man, has significance for the clinical behavior of germ cell tumors. Another observation, pertaining to the induction of certain glycolipid antigens on differentiated teratocarcinoma cells by human cytomegalovirus (HCMV) and human immunodeficiency virus (HIV), provides insights into the mechanisms whereby these viruses might damage the developing embryo. The biological and clinical implications of these cell surface markers of human teratocarcinoma cells will be discussed.

## Cell Surface Marker Antigens of Human Embryonal Carcinoma Cells

### *Glycolipid Antigens*

Human EC cells commonly express two monoclonal antibody-defined embryonic antigens known as stage-specific embryonic antigens 3 and 4 (SSEA-3 and SSEA-4) (Andrews et al. 1982, 1984b). Both antigens are expressed on cleavage stage mouse embryos but not on murine EC cells or the later primitive ectoderm to which murine EC cells are thought to correspond (Shevinsky et al. 1982; Kannagi et al. 1983a,b). These and other circumstantial data (e.g., the common appearance of trophoblastic giant cells in human but not murine testicular or embryo-derived teratocarcinomas) led us to suggest (Andrews et al. 1980, 1982) that human EC cells correspond to an earlier embryonic cell type than do murine EC cells. However, biochemical considerations cast doubt on the basis for this hypothesis and, indeed, suggest more fundamental similarities between murine and human EC cells than were at first apparent.

The antibodies that define both SSEA-3 and SSEA-4 were shown to recognize epitopes carried by cell surface glycolipids of the globoseries (Table 1) (Shevinsky et al. 1982; Kannagi et al. 1983a,b). Specifically, SSEA-3 was identified as the terminal trisaccharide of an extended globoseries lipid, Galβ1→3Globoside (GL5), and SSEA-4 as its sialated derivative, NeuNAcα2→3Galβ1→3Globoside (GL7). Glycolipid analyses further revealed that globoseries glycolipids, in contrast to lacto- and ganglioseries structures, predominate in both murine and human EC cells (Fenderson et al. 1987; Willison et al. 1982; Kannagi et al. 1983a). However, in murine EC cells terminal modification of globoside occurs by the addition of GalNac, rather than Gal, resulting in the expression of the Forssman antigen rather than SSEA-3 and SSEA-4. Thus, the distinction between murine and human EC cells based on differential expression of these antigens should be interpreted cautiously. It may well reflect species differences in the capacity for particular terminal modifications of oligosaccharide core structures and not necessarily differences in the developmental stage of the cells to which these embryo-like neoplastic cells correspond.

**Table 1.** Glycolipid antigens of NTERA-2 EC cells and their differentiated derivatives

| Antibody | Antigen/glycolipid | Glycolipid structure | Reference |
|---|---|---|---|
| | | *Globoseries* | |
| | Pᵏ, GL3 | Galα1→4Galβ1→4Glcβ1→Cer | Naiki and Marcus (1974) |
| | P, globoside | GalNAcβ1→3Galα1→4Galβ1→4Glcβ1→Cer | Naiki and Marcus (1974) |
| MC631 | SSEA-3 | Galβ1→3GalNAcβ1→3Galα1→4Galβ1→4Glcβ1→Cer | Shevinsky et al. (1982) |
| MC813–70 | SSEA-4 and SSEA-3 | NeuAcα2→3Galβ1→3GalNAcβ1→3Galα1→4Galβ1→4Glcβ1→Cer | Kannagi et al. (1983a) |
| MBr1 | Globo-H | Fucα1→2Galβ1→3GalNAcβ1→3Galα1→4Galβ1→4Glcβ1→Cer | Bremer et al. (1984) |
| HH5 | Globo-A | Fucα1→2Galβ1→3GalNAcβ1→3Galα1→4Galβ1→4Glcβ1→Cer<br>3<br>↑<br>GalNAcα1 | Clausen et al. (1986) |
| | | *Lactoseries* | |
| MC480 | SSEA-1, $Le^x$ | Galβ1→4GlcNAcβ1→3Galβ1→4Glcβ1→Cer<br>3<br>↑<br>Fucα1 | Solter and Knowles (1978)<br>Kannagi et al. (1982) |
| AH6 | $Le^y$ | Galβ1→4GlcNAcβ1→3Galβ1→4Glcβ1→Cer<br>2 3<br>↑ ↑<br>Fucα1 Fucα1 | Abe et al. (1983) |
| | | *Ganglioseries* | |
| VIN-IS-56 | GD3 | NeuNAcα2→8NeuNAcα2→3Galβ1→4Glcβ1→Cer | Andrews et al. (1990) |
| VIN-2PB-22 | GD2 | NeuNAcα2→8NeuNAcα2→3Galβ1→4Glcβ1→Cer<br>4<br>↑<br>GalNAcβ1 | Andrews et al. (1990) |
| ME311 | 9-*O*-acetyl-GD3 | 9-*O*-acetylNeuNAcα2→8NeuNAcα2→3Galβ1→4Glcβ1→Cer | Thurin et al. (1985) |
| A2B5 | GT3 | NeuNAcα2→8NeuNAcα2→8NeuNACα2→3Galβ1→4Glcβ1→Cer | Eisenbarth et al. (1979)<br>Fenderson et al. (1987) |

[a] This table only lists antibodies that we have used extensively in our studies. Further, although the specificities shown for the various antibodies are the principal reactivities relevant to the NTERA-2 system (Fenderson et al. 1987; Andrews et al. 1990), the list is necessarily incomplete and the reader is referred to the cited references for fuller details. For example, although A2B5 has been reported to react with various polysialogangliosides, GT3 appears to account for most of the A2B5 reactivity in differentiated NTERA-2 cells.

In human EC cells, other terminal modifications of globoseries glycolipids are also found, notably modifications leading to the expression of ABH blood group epitopes – the so-called globo-ABH antigens (Fenderson et al. 1987). In the adult, ABH antigens are mainly associated with lactoseries glycolipids – lacto-ABH antigens (Hakomori 1981). However, the polymorphism of the globo-ABH and lacto-ABH antigens seems to be under the same genetic control: both globo-A and globo-H reactive glycolipids are expressed by TERA 2 EC cells, which were derived from a blood group A patient, whereas EC cells from a blood group O patient express only globo-H (Fenderson et al. 1987). We speculated that the globoseries blood group antigens are predominantly embryonic and that a switch in the synthesis of glycolipid oligosaccharide core structures from globo- to lactoseries occurs during the later stages of human development. Such a switch is evident during the differentiation of pluripotent TERA 2 EC cells and is discussed below (p. 74).

The globoseries glycolipids also carry the epitopes of the P blood group system (Race and Sanger 1975). In particular, the common red cell antigen P, which is expressed by almost all individuals, has been associated with globoside itself (Table 1) (Naiki and Marcus 1974). Likewise, SSEA-3 and SSEA-4 are also expressed on the red cells of most people (Tippett et al. 1986). However, very rare individuals appear to lack the enzymes required for synthesis of the globoseries core oligosaccharide; they either express no antigens of this series, including SSEA-3 or SSEA-4, on their red cells (p phenotype), or they express another antigen ($P^k$) associated with the Galα1→4Galβ1→4Glcβ1→ceramide trisaccharide core structure, but not the P antigens, or SSEA-3 or SSEA-4 which are all associated with longer oligosaccharides (Table 1).

The p and $P^k$ phenotypes are due to recessive genetic traits, and high rates of spontaneous abortion occur in p and $P^k$ women, apparently due to an immune response against embryonic antigens genetically lacking in the mother (Levine 1977; Soderstrom et al. 1984; Lopez et al. 1983; Bono et al. 1981). From our studies of human EC cells it seems likely that the globoseries antigens SSEA-3 and SSEA-4 are expressed on the early human embryo and are the targets of the immune response in p and $P^k$ women.

Whereas p and $P^k$ individuals are exceedingly rare, another polymorphic variant belonging to the P system, luke(−) (Tippett et al. 1965), is rather more common and is found in about 2% of the population. In a more recent study (Tippett et al. 1986), we found that Luke(−) individuals also did not express SSEA-4 on their red cells, although they were SSEA-3$^+$; conversely, SSEA-4$^-$ individuals were also Luke(−). The results suggested an identity between the Luke antigen and SSEA-4, and that the genetic defect in Luke(−) individuals is an inability to add sialic acid to the extended globoseries carbohydrate core structure. Since SSEA-4 is probably expressed by early human embryos, one might wonder whether Luke(−)/SSEA-4$^-$ women would exhibit the same high rates of spontaneous abortion as p and $P^k$

women. With about 1%–2% of the population being of this phenotype, this might prove to be a significant cause of habitual abortions.

Another issue is the possible significance of P-blood group polymorphism for understanding the clinical behavior of germ cell tumors. Whereas SSEA-3 and SSEA-4 are commonly expressed by human EC cells, this is evidently not a universal property. Of particular note are several SSEA-3$^-$/SSEA-4$^-$ EC clones derived from the TERA 2 line (Andrews et al. 1985; Thompson et al. 1984). Glycolipid analyses of one of these revealed that globoseries glycolipids are as predominant in this line as in SSEA-3$^+$/SSEA-4$^+$ clones, but that the SSEA-3 and SSEA-4 epitopes are apparently hidden by an, as yet, undetermined terminal modification of Gal-globoside (Fenderson et al. 1987). More striking was another observation that this SSEA-3$^-$/SSEA-4$^-$ clone converted to an SSEA-3$^+$/SSEA-4$^+$ phenotype when grown as a xenograft in a nude mouse (Andrews et al. 1985); the SSEA-3$^+$/SSEA-4$^+$ phenotype persisted when the tumors were explanted and sublines of EC cells grown out in vitro. While the question of cause or effect remains to be resolved, this observation raises the question of what role the globoseries antigens play in tumor growth, and whether the genotype of germ cell tumor patients with respect to the P blood group system might affect the behavior of their tumors – for example, would EC tumors arising in Luke(−)/SSEA-4$^-$ patients show the same patterns of growth and differentiation as EC tumors in Luke(+)/SSEA-4$^+$ patients.

### *(Glyco)protein Antigens*

With the exception of the highly glycosylated mucin-like proteins and proteoglycans, which we discuss separately below (p. 68), the two groups of surface glycoproteins that have received the most attention in studies of human germ cell tumors are the alkaline phosphatases (ALP) and the major histocompatibility complex (MHC) antigens. More recently the expression growth factor receptors has also been studied (e.g., Carlin and Andrews 1985; Engström et al. 1985; Weima et al. 1988), but this will be discussed in more detail by other contributors to this volume.

Most human EC cell lines that have been studied express high levels of total ALP activity, much of which is heat labile and has been shown to correspond to the liver/bone/kidney form of the enzyme (L-ALP) (Benham et al. 1981). ALP is found on the cell surface and is readily detectable immunologically. For example, two monoclonal antibodies that we raised to 2102Ep human EC cells, TRA-2-49 and TRA-2-54, were shown to be specific for L-ALP and to provide useful additional markers of human EC cells (Andrews et al. 1984c). These results correlate well with observations in murine EC cells which also express high levels of L-ALP activity (Hass et al. 1979).

Clinically, however, the placental, or more correctly placental-like, isozymes of ALP, P-ALP, have made more impact. These variants are commonly expressed ectopically by a variety of tumors but have attracted attention in the context of germ cell tumors because they are strongly expressed by seminomas (Hustin et al. 1987; Uchida et al. 1981; Wahren et al. 1979). P-ALP has been proposed as both a histochemical marker and a serum marker, useful in distinguishing seminomatous and nonseminomatous germ cell tumors. Nevertheless, we and others have consistently observed P-ALP in cell lines apparently composed of EC cells, in which this isozyme generally accounts for a few percent of the total ALP activity (Andrews et al. 1980, 1981; Benham et al. 1981; Cotte et al. 1981), and in tumor biopsies from patients with EC (Paiva et al. 1983). It is , however, unclear whether this P-ALP activity is attributable to the EC cells themselves or to some spontaneous differentiated derivatives.

Unlike murine EC cells, many human EC cell lines express low levels of the MHC class I, but not class II, antigens (Andrews et al. 1980, 1981, 1982). Nevertheless, in some lines, TERA 2 EC clones being a good example, the low expression is quite variable and may not be detected (e.g., see Thompson et al. 1984; Andrews et al. 1984b, 1985). The reasons for this variability are unclear, but, as we have discussed previously, these low levels of MHC expression apparently do not indicate the presence of differentiated cells, as they do in the murine teratocarcinomas (Andrews et al. 1981, 1983b).

The expression of MHC antigens is modulated by exposure to interferon (IFN), notably IFN-$\gamma$ (Andrews et al. 1987). As in many other tumor cells, IFN-$\gamma$ exposure results in the increased expression of MHC class I, although not class II, antigens in both 2102Ep and TERA 2 human EC cells, and in the 1411H human yolk sac carcinoma cell line. However, this is not accompanied by any concomitant differentiation. Interestingly, some murine EC cells can be induced to express MHC antigens by exposure to IFN-$\gamma$ (Andrews et al. 1987; Wan et al. 1987), indicating that the differences between human and murine EC cells with regard to MHC expression are more quantitative than qualitative. In the EC cells of both species, the response to IFN is only partial since it is not associated with a full range of protection to viral infection in either case (Andrews et al. 1987; Wood and Hovanessian 1979).

### *Mucin-like Antigens*

In common with a variety of other tumor cells, EC cells express high molecular weight, highly glycosylated proteins – proteoglycans and mucin-like molecules (e.g., Muramatsu et al. 1979, 1982; Rettig et al. 1985; Pera et al. 1988). Among the monoclonal antibodies that we have raised to human EC cells, two, TRA-1-60 and TRA-1-81, also appear to recognize epitopes associated with mucin-like molecules (Andrews et al. 1984a). Both antigens are strongly expressed by human EC cells and are useful supplements to

SSEA-3 and SSEA-4 for identifying these cells in vitro. Further, as we shall discuss below, the TRA-1-60 and TRA-1-81 antigens offer the prospect of new serum markers for monitoring therapy of germ cell tumor patients.

From lysates of human EC cells, surface labeled with $^{125}I$, both TRA-1-60 and TRA-1-81 immunoprecipitate a pair of polypeptides with apparent molecular weights of approximately 200000 and 400000 (Andrews et al. 1984a). However, cross-clearing experiments indicated that the epitopes are carried by different polypeptides, suggesting a family of antigenically distinct, but structurally similar molecules. Further experiments indicated that the cell surface reactivity with antibody TRA-1-60, but not TRA-1-81, could be eliminated by digestion with neuraminidase, implying that the TRA-1-60 epitope involves a sialic acid residue (Table 2). More recently, we analyzed soluble antigen in cell lysates or culture supernatants by its ability to inhibit antibody binding to fixed target cells, and we have found that both the TRA-1-60 and TRA-1-81 epitopes are sensitive to periodate treatment and to digestion with alkaline borohydride (Table 3). These observations are consistent with the hypothesis that both epitopes are carried by 'O'-linked oligosaccharide side chains of a polypeptide backbone, a structure typical of mucins. The probability that both TRA-1-60 and TRA-1-81 epitopes are associated with mucin-like molecules is further strengthened by their relatively high density on CsCl isopyknic centrifugation (Fig. 1).

Typical of tumor mucin-like antigens, both TRA-1-60- and TRA-1-81-reactive molecules can be detected in the culture supernatants of human EC cells growing in vitro. This suggested that they might also be detectable in the serum of germ cell tumor patients, and supplement the serum tumor markers

**Table 2.** Effect of neuraminidase on the cell surface reactivity of monoclonal antibodies TRA-1-60 and TRA-1-81

| | Fluorescence intensity (arbitary units) | | | |
|---|---|---|---|---|
| | P3X | MC813-70 (anti-SSEA-4) | TRA-1-60 | TRA-1-81 |
| No enzyme | 4 | 128 | 176 | 94 |
| Neuraminidase (VC)[a] | 8 | 13 | 15 | 156 |
| Neuraminidase (AU)[a] | 7 | 101 | 11 | 165 |

[a] 2102Ep human EC cells were incubated for 30 min at 37°C in phosphate-buffered saline containing 0.1 U/ml neuraminidase from either *Vibrio cholera* (VC) or *Arthrobacter ureafaciens* (AU), obtained from Calbiochem. VC neuraminidase hydrolyzes $\alpha2\rightarrow3$, $\alpha2\rightarrow6$, and $\alpha2\rightarrow8$ linkages; AU neuraminidase is specific for $\alpha2\rightarrow6$ linkages. The binding of TRA-1-60 and TRA-1-81 antibodies to such neuraminidase-treated cells was determined by an immunofluorescence assay using a flow cytofluorimeter; mean fluorescence intensity was measured on an arbitrary scale from 1 to 200. Negative control antibody P3X and antibody MC813-70, which detects the sialated globoseries glycolipid GL7 (Table 1), were included as controls.

**Table 3.** Stability of soluble preparations of TRA-1-60 and TRA-1-81 antigens to various chemical and enzymatic treatments

| | % inhibition | |
|---|---|---|
| Treatment | TRA-1-60 | TRA-1-81 |
| None | 83 | 82 |
| 37°C, 24 h | 59 | 69 |
| 10 mg/ml pronase, 37°C, 24 h | 94 | 73 |
| 50 m*M* periodate, 37°C, 24 h | 0 | 0 |
| 1*M* $NaBH_4$ in 0.01*N* NaOH, 37°C, 24 h | 0 | 0 |

2102Ep human EC cells were lysed with 0.75% octylthioglycoside; after clarification by centrifugation at 10 000 *g* for 25 min, the presence of soluble antigen was assayed by its ability to inhibit a standard radioimmunobinding assay to glutaraldehyde-fixed 2102Ep target cells. Approximately 50% inhibition was achieved with a lysate containing the equivalent of $5 \times 10^6$ cells/ml. Note that both periodate and alkaline borohydride digestions resulted in the removal of all detectable TRA-1-60- and TRA-1-81-inhibiting activity from the lysates, whereas the activity was stable when exposed to proteolytic digestion.

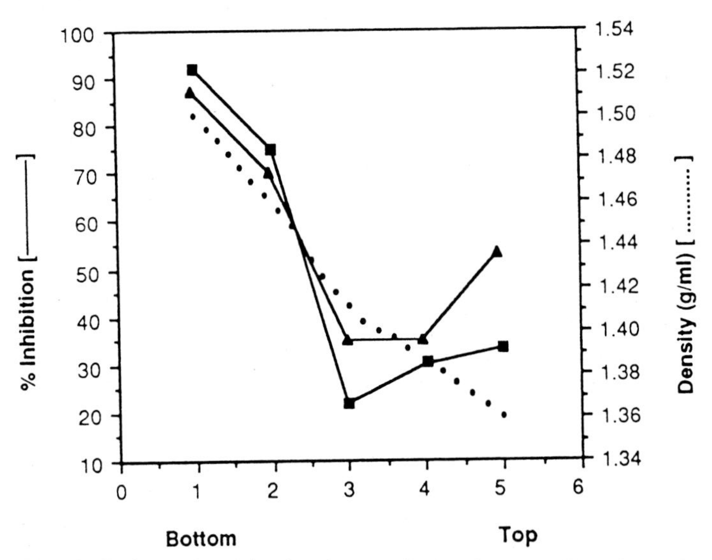


**Fig. 1.** Cesium chloride density gradient ultracentrifugation of soluble TRA-1-60 (▲-▲) and TRA-1-81 (■-■) antigens, partially purified from 2102Ep EC cell culture supernatant by gel filtration. Density was evaluated by refractive index measurement, and antigenic reactivity was determined by inhibition of a radioimmunobinding assay to fixed target cells

in current clinical use, such as human chorionic gonadotropin (HCG) and $\alpha$-fetoprotein (AFP). As we and others have established (Suurmeijer et al. 1983), AFP and HCG are produced by yolk sac and choriocarcinomata, respectively, and are found in only about 80% of patients with nonseminomatous germ cell tumor (NSGCT). Moreover, because of the histological complexity of these tumors, the longitudinal profiles of HCG and AFP seen during the follow-up of patients initially positive for these markers do not necessarily parallel the clinical outcomes, since non(AFP/HCG)-producing tumor cells sometimes survive the chemotherapeutic regimen. Being products of EC cells, TRA-1-60 and TRA-1-81 antigens could be used even in patients whose tumors do not show any yolk sac or trophoblastic differentiation, presumably the 20% of NSGCTs that are not associated with elevated serum AFP and HCG.

Using an extract of TERA 2 EC cells as the antigen source, we have developed an ImmunoEnzymoMetricAssay for the detection of TRA-1-60-reactive antigens in serum. Briefly, antigen is incubated with excess antibody, after which the antigen/antibody complex is precipitated by 2.5% polyethylene glycol and the nonbound antibodies in the supernatant are transferred to antigen-coated wells of a microtiter plate. The amount of unbound antibody is then quantified by its ability to bind to the microtiter wells as detected by a peroxidase-coupled second antibody. The bound peroxidase activity is assayed with orthophenylene diamine as a substrate, the amount of product being measured by extinction at 492 nm. Thus, in this indirect assay the extinction measured is inversely related to the amount of antigen present in the original sample, which is quantified in arbitrary units per milliliter (U/ml). After optimization of this assay, we tested the sera of our target group of nonseminomatous testicular tumor patients (NSTTs), who were histologically classified for the presence or absence of EC elements.

Serum from 38 normal control individual and 19 patients with pure seminoma did not contain detectable TRA-1-60 antigen (i.e., $<300$ Uml). On the other hand, the TRA-1-60 marker was detected at levels of 300–13 000 U/ml in sera from 14 of 15 EC-positive patients. In the sera of seven NSGCT patients whose tumors were devoid of EC, TRA-1-60 was elevated in only one case (1300 U/ml). It remains to be determined whether this exception in the latter group was due to an EC element in the tumor, which was overlooked during histological examination.

In this pilot study, the TRA-1-60 profiles paralleled the AFP curves in 13 out of the 15 patients monitored. In one patient who was negative for AFP, the TRA-1-60 curve followed the profile of the HCG curve. Moreover, in another patient who was negative for both AFP and HCG, TRA-1-60 appeared to be the only positive marker. Though in general TRA-1-60 seems to follow the AFP/HCG curves, apparently depending on the type of tumor cell remaining after chemotherapy, the three markers might diverge one from another during follow-up as shown in three representative examples (Fig. 2). Thus, TRA-1-60-reactive antigen has the potential for clinical application as

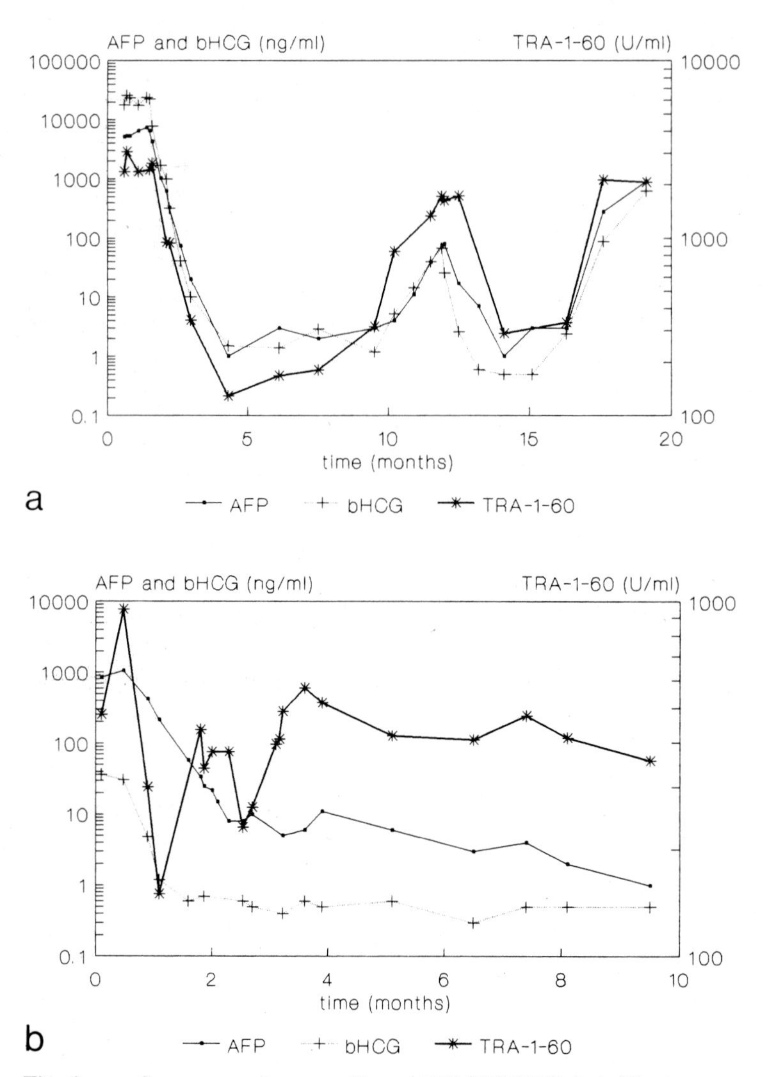


**Fig. 2a–c.** Serum marker profiles (AFP/HCG/TRA-1-60) in nonseminomatous testicular tumor patients. **a** Follow-up after institution of chemotherapy in a patient with NSTT stage III. The primary tumor contained EC and mature teratoma. **b** Marker profiles in a patient with stage III EC. Note that AFP/HCG and TRA-1-60 markers only parallel each other in the early stages of follow-up, but later they behave differently. **c** See p. 73

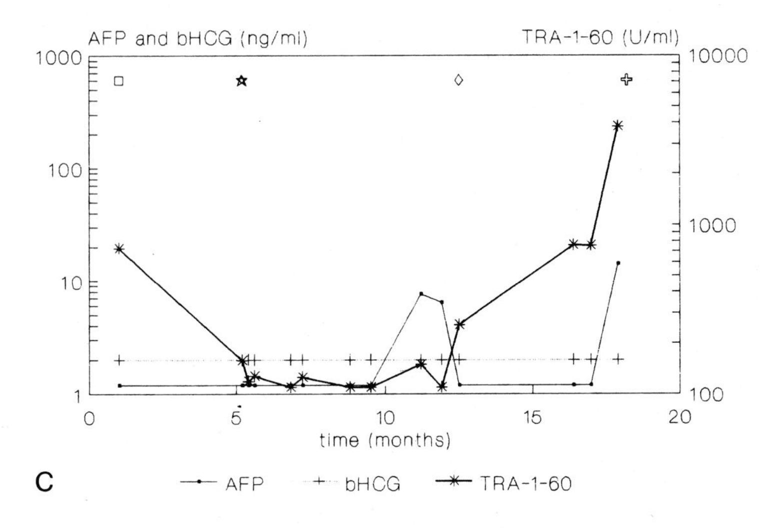


**Fig. 2** (*continued*). **c** Marker profiles and clinical data in an NSTT patient with EC carcinoma in his primary tumor. Note that, whereas AFP and HCG remained within their normal ranges during follow-up, the reappearance of serum TRA-1-60 clearly presaged the subsequent death of the patient

an additional marker in NSGCT($EC^+$) patients and deserves to be studied in greater detail. Our preliminary data on the assay procedure and the individual patient data will be published more extensively elsewhere.

## Differentiation

### *Induction of Differentiation*

Almost all of our own studies of human EC cell differentiation have been confined to clones of the human testicular teratocarcinoma-derived lines 2102Ep (Andrews et al. 1982) and TERA 2 (Andrews et al. 1984b). Another series of TERA 2 clones have been derived and studied independently by Graham and his colleagues with broadly similar results (Thompson et al. 1984). Several other pluripotent human EC cell lines have now been reported (e.g., Sekiya et al. 1985; Teshima et al. 1988; Pera et al. 1989) and are discussed by other contributors to this symposium.

Like many other human EC cells, 2102Ep cells do not differentiate in response to retinoic acid (Matthaei et al. 1983), a potent inducer of differentiation of many murine EC cells (Strickland and Mahdavi 1978) which has recently been shown to act as an embryonic morphogen (Thaller and Eichele 1987). Nevertheless, when grown at low cell densities, a proportion of 2102Ep

EC cells undergo changes that appear to constitute cellular differentiation: the cells flatten out and enlarge, and up to 50% or more begin expressing the lactoseries glycolipid antigen SSEA-1, which is not usually expressed by undifferentiated human EC cells (Andrews et al. 1982). Further, synthesis of fibronectin is induced in the SSEA-$1^+$ cells (Andrews 1982), while a few cells acquire an ultrastructure consistent with differentiation into trophoblastic giant cells (Damjanov and Andrews 1983). Indeed, a small number of cells in these low-density cultures were shown by immunohistochemistry to contain cytopplasmic HCG. Nevertheless, this limited capacity for differentiation, apparently along a trophoblastic lineage, is incomplete and many EC cells persist and overgrow the cultures. Furthermore, the differentiation is not readily controlled and is quite variable from experiment to experiment. Thus, the 2102Ep cell line is not the most suitable model for studying the differentiation of human EC cells.

On the other hand, TERA 2 EC cells, especially the series of NTERA-2 clones that we have studied, can be induced to differentiate in ways that are more amenable to experimental study (Andrews et al. 1984b; Thompson et al. 1984). These EC cells form well-differentiated tumors in nude mice, but, more conveniently, they differentiate almost completely in vitro when exposed to retinoic acid (Andrews 1984), hexamethylene bisacetamide (HMBA), or 5-bromouracil deoxyribose (BUdR) (Andrews et al. 1986, 1990). Retinoic acid-induced differentiation leads to a heterogeneous population of cells among which the most striking are neurons. These neurons express cell surface tetanus toxin receptors (Andrews 1984), all three neurofilament proteins (Lee and Andrews 1985), and S100b protein (Tsutsui et al. 1987). Moreover, they exhibit ion channels and regenerative membrane potentials resembling those of embryonic neurons (Rendt et al. 1989). Other changes that occur upon differentiation include the appearance of susceptibility to productive infection with HCMV (Gönczöl et al. 1984) and HIV (Hirka et al. 1991), and the expression of a number of homeobox-containing genes (Mavilio et al. 1988).

The differentiation induced by BUdR resembles that induced by retinoic acid, and also leads to the formation of prominent neurons (Andrews et al. 1986, 1990). On the other hand, the cultures induced by HMBA are morphologically and antigenically quite disctinct from either the retinoic-acid- or BUdR-treated cultures, and contain very few neurons (Andrews et al. 1986, 1990). Thus, TERA 2 EC cells appear capable of differentiating along at least two distinct pathways, and this process can be manipulated by the chemical nature of the stimulus employed.

### *Antigenic Changes upon Differentiation*

Within 7–14 days of first exposure to retinoic acid ($10^{-5}$ *M*), the globoseries antigens SSEA-3 and SSEA-4 disappear from the surface of TERA 2 EC cells, correlating with the disappearance of an EC cell morphology (Fenderson et al. 1987). Similarly, there is a marked reduction in globoseries glycolipids as shown by thin-layer chromatography of total lipid extracts of the cells. Interestingly, surface reactivity of anti-SSEA-3 disappears more quickly than surface reactivity of anti-SSEA-4, even though both monoclonal antibodies bind to the sialated globoseries lipid GL7. This paradoxical result seems to be due to nonspecific inhibition of anti-SSEA-3 binding by gangliosides, especially $GD_3$, which appear during retinoic-acid-induced differentiation. Indeed, this differentiation is marked by a shift in the synthesis of glycolipid oligosaccharide core structures from globo- to lacto- and ganglioseries cores and by the corresponding appearance of several new lacto- and ganglioseries cell surface antigens (Fenderson et al. 1987). The key regulatory enzymes in this switch appear to be α-galactosyltransferase, which initiates synthesis of globo-core structures from the precursor, lactosyl ceramide, and is reduced in activity upon differentiation, and β-*N*-acetylglucosaminyl and sialyl transferases, which initiate synthesis of lacto- and ganglio-core structures, respectively, and are increased in activity upon differentiation (Chen et al. 1989).

The expression of several of the EC cell glycoprotein antigens also changes upon the retinoic-acid-induced differentiation of TERA 2 EC cells. For example, the L-ALP-related antigens decrease as does Thy-1, which is strongly expressed on the stem cells (Andrews et al. 1983a). The levels of MHC class I antigens either do not change significantly during this process or are even reduced. However, exposure of the differentiated cells to IFN-γ results in much higher levels of MHC antigen expression than in the IFN-γ-treated EC cells (Andrews et al. 1987). Nevertheless, the most striking change affecting glycoprotein determinants is the disappearance of the TRA-1-60 and TRA-1-81 mucin-like antigens; strongly expressed by the EC cells, these antigens disappear from most cells within 14 days, although they persist on a small subset of differentiated cells.

The differentiated cells induced by retinoic acid are clearly heterogeneous, and subsets of cells can be discerned by their patterns of surface antigen expression. Many of the cells express ganglioseries antigens such as A2B5, ME311, VIN-IS-56, and VIN2PB22, and subsets can be found expressing one but not another of these antigens. In particular, almost all the neurons express A2B5, VIN-IS-56, and VIN2PB22 antigens, but they lack the ME311 antigen (Fenderson et al. 1987; Andrews et al. 1990). The $ME311^{+}$ cells do not resemble neurons and do not express neurofilament proteins.

Islands of large flat cells, morphologically distinct from the surrounding differentiated cells, constitute another subset. The cells of these islands express lactoseries antigens, notably SSEA-1, but not the ganglioseries antigens such as A2B5 which are predominant on the surrounding cells (Andrews 1988). They also continue to express the TRA-1-60 and TRA-1-81 antigens. Interestingly, prominent glandular structures present in xenograft tumors of TERA 2 also express SSEA-1 as well as TRA-1-60 and TRA-1-81. We have speculated that the $SSEA\text{-}1^{+}/TRA\text{-}1\text{-}60^{+}/TRA\text{-}1\text{-}81^{+}$ islands in the retinoic-acid-induced cultures represent an attempt to form the same glandular structures in vitro.

Whereas BUdR and retinoic acid induce similar cell surface antigen changes, HMBA-induced differentiation is distinct. The EC marker antigens disappear with similar kinetics following treatment with all three inducers, but the ganglioseries and lactoseries antigens induced by retinoic acid do not appear on HMBA-induced cells, at least during the first 14 days of treatment (Andrews et al. 1990). At present the only surface marker that we have found to be induced by HMBA is an antigen detected by a new monoclonal antibody, VIN-IS-53, raised against retinoic-acid-induced, differentiated TERA 2 cells. Eventually, after 14 days exposure to HMBA, a few cells expressing antigens common in retinoic-acid-treated cultures do begin to appear, and neurons are occasionally detected. Consequently, while retinoic acid and HMBA appear to promote differentiation along predominantly different lineages, the distinction is really quantitative and there is some overlap in the types of cells induced.

### *Effects of Infection with HCMV and HIV*

Whereas TERA 2 EC cells are not permissive for the infection and replication of either HCMV or HIV, many of their differentiated derivatives induced by retinoic acid or HMBA are permissive for the full replication of both viruses (Gönczöl et al. 1984; Andrews et al. 1986; Hirka et al. 1990). In the case of HCMV, the block to replication in the EC cells appears to be at the level of transcription of the immediate early genes (Lafemina and Hayward 1986; Nelson and Groudine 1986), but in the case of HIV a block at the level of RNA processing rather than transcription has been proposed (Maio and Brown 1988).

While attempting to identify the cell surface antigen phenotype of the HCMV-susceptible cells in differentiated TERA 2 cultures, we discovered that HCMV-infected cells commonly express the lactoseries antigen SSEA-1 (Fig. 3) (Andrews et al. 1989). Further, it appeared that this was not due to preferential infection of SSEA-$1^+$ cells, but that HCMV infection actually induces expression of this antigen. This conclusion was confirmed by studies in human embryo fibroblasts which did not express SSEA-1 under any growth conditions tested unless infected with HCMV (Fig. 3). The mechanism for induction is not simple, but appears to involve multiple effects on the activities of several glycosyltransferases, the net result of which is the appearance of SSEA-1-reactive glycolipids on the cell surface. Nevertheless, one key change leading to this result might be the increased activity of β-*N*-acetylglucosaminyl transferase in the infected cells, since this is though to govern the rate-limiting step in the synthesis of the lactoseries glycolipid core structure.

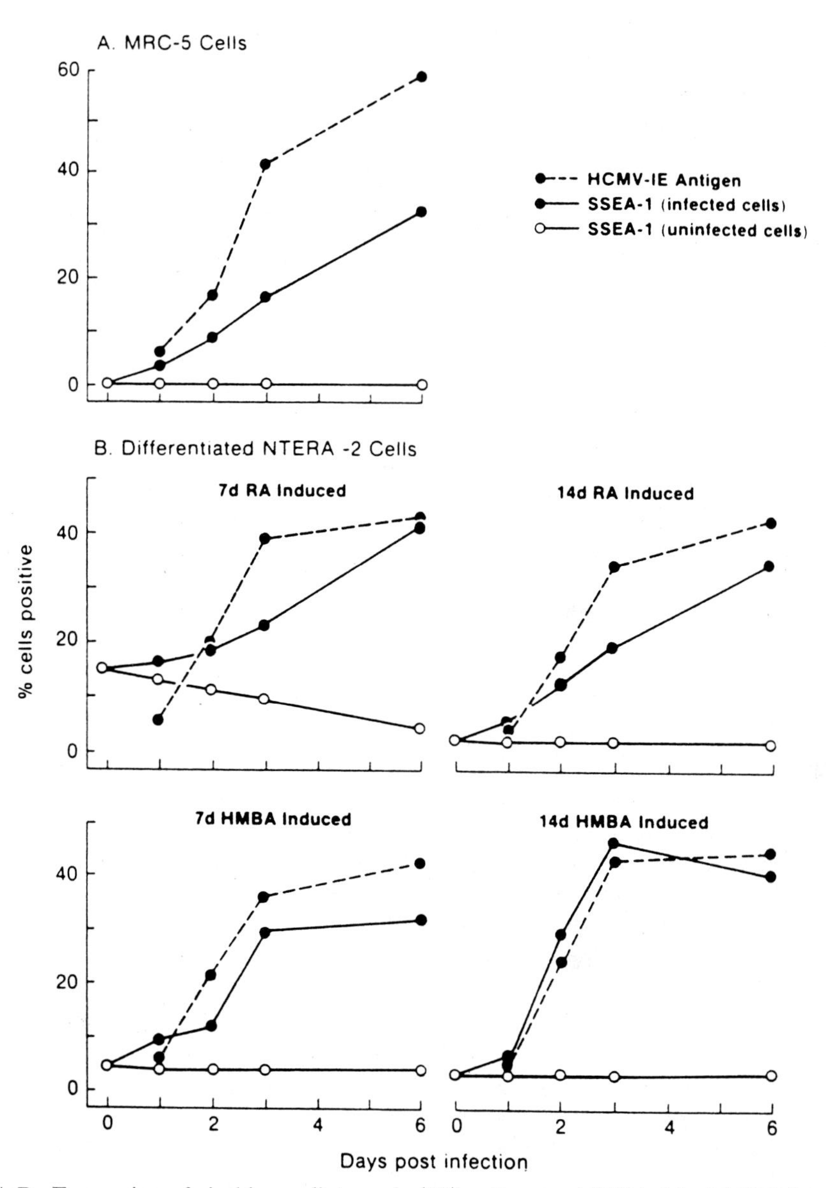


**Fig. 3A,B.** Expression of viral immediate early (*IE*) antigen and SSEA-1 in **A** MRC-5 fibroblasts and **B** differentiated NTERA-2 cells following infection with HCMV. The antigen expression was determined by immunofluorescence. Differentiated NTERA-2 cells were derived by prior exposure to either retinoic acid (*RA*) or HMBA for 7 or 14 days. Note the induction of SSEA-1 following infection. Although some SSEA-1$^+$ cells are found in RA-induced cultures in the absence of infection, few if any are found in HMBA-induced cultures or in MRC-5 fibroblasts without infection. In the latter cases the SSEA-1$^+$ cells can also be shown to be HCMV infected, using two-color immunofluorescence. (See Andrews et al. 1989)

Like HCMV, we have also observed that HIV induces lactoseries antigens in infected NTERA-2 cells, although in this case the predominant epitope detected is the $Le^y$ structure rather than the related $Le^x$ structure, the epitope recognized as SSEA-1 (see Table 1) (Hirka et al. 1991). A similar observation has also been made in HIV-infected lymphocytes by Adachi et al. (1988).

Intrauterine infection with HCMV often leads to damage to the developing embryo and especially to the developing nervous system (Stagno et al. 1983). Similarly, HIV infection has also been suggested as a potential cause of birth defects (e.g., Marion et al. 1986). In this light it is of interest to note that lactoseries glycolipids, especially the $Le^x$ structure or SSEA-1, have been suggested to play a role in mediating critical cell-cell interactions during embryonic development, first during compaction of the morula (Bird and Kimber, 1984; Fenderson et al. 1984) and later in the developing nervous system (Dodd and Jessel 1985, 1986). The question is raised, therefore, whether the inappropriate expression of $Le^y$ or $Le^x$ in time or space, induced by viral infection, might contribute to the teratogenic effects attributed to HCMV or possibly HIV.

## Conclusion

Our studies of human EC cell lines in vitro confirm the value of surface antigens, especially glycolipids and mucins, as markers, both for the identification of EC cells and for following their differentiation. Our data suggest that some of the markers could serve as aids in clinical pathology to distinguish the various subtypes of germ cell tumors. Further, the EC mucin antigens defined by monoclonal antibodies TRA-1-60 and TRA-1-81 may provide valuable new serum markers for monitoring germ cell tumor therapy, especially in patients whose tumors do not produce HCG and AFP.

The function of these cell surface molecules during embryonic development or during oncogenesis remains to be established. Nevertheless, we have alluded to clues to their possible clinical significance – for example, the involvement of P-blood group polymorphism in habitual abortion and the potential for its effect on the behavior of germ cell tumors. Finally, the induction of lactoseries antigens might be involved in the teratogenicity of certain viruses.

*Acknowledgements.* This work was supported by United States Public Health Service grants CA29894 and AI24943 from the National Institutes of Health. Alexander von Keitz was a recipient of a fellowship from the Deutsche Forschungsgemeinschaft. We are grateful for the excellent technical assistance of Sandra Kumpf, Leslie Marinelli, Kyle Wagner, Petra van Brummen, and Rina de Jong.

## References

Abe K, McKibbin JM, Hakomori S (1983) The monoclonal antibody directed to difucosylated type II chain (Fucα1→2Galβ1→4[Fucα1→3] GlcNAc; Y determinant). J Biol Chem 258:11793–11797
Adachi M, Hayami M, Kashiwagi N, Mizuta T, Ohta Y, Gill MJ, Matheson DS, Tamaoki T, Shiozawa C, Hakomori S-I (1988) Expression of $Le^y$ antigen in human immunodeficiency virus-infected human T cell lines and in peripheral lymphocytes of patients with acquired immunodeficiency syndrome (AIDS) or AIDS-related complex (ARC). J Exp Med 167:323–331
Andrews, PW (1984) Retinoic acid induces neuronal differentiation of a cloned human embryonal carcinoma cell line in vitro. Dev Biol 103:285–293
Andrews PW (1988) Human teratocarcinoma. Biochim Biophys Acta 948:17–36
Andrews PW, Bronson DL, Benham F, Strickland S, Knowles BB (1980) A comparative study of eight cell lines derived from human testicular teratocarcinoma. Int J Cancer 26:269–280
Andrews PW, Bronson DL, Wiles MV, Goodfellow PN (1981) The expression of major histocompatibility antigens by human teratocarcinoma derived cell lines. Tissue Antigens 17:493–500
Andrews, PW, (1982) Human embryonal carcinoma cells in culture do not synthesize fibronectin until they differentiate. Int J Cancer 30:567–571
Andrews PW, Goodfellow PN, Shevinsky L, Bronson DL, Knowles BB (1982) Cell surface antigens of a clonal human embryonal carcinoma cell line: morphological and antigenic differentiation in culture. Int J Cancer 29:523–531
Andrews PW, Goodfellow PN, Bronson DL (1983a) Cell surface characteristics and other markers of differentiation of human teratocarcinomas in culture. Cold Spring Harbor Conf Cell Proliferation 10:579–590
Andrews PW, Goodfellow PN, Damjanov, I (1983b) Human teratocarcinoma cells in culture. Cancer Surv 2:41–73
Andrews PW, Banting GS, Damjanov I, Arnaud D, Avner P (1984a) Three monoclonal antibodies defining distinct differentiation antigens associated with different high molecular weight polypeptides on the surface of human embryonal carcinoma cells. Hybridoma 3:347–361
Andrews PW, Damjanov I, Simon D, Banting G, Carlin C, Dracopoli NC, Fogh J (1984b) Pluripotent embryonal carcinoma clones derived from the human teratocarcinoma cell line TERA-2: differentiation in vivo and in vitro Lab Invest 50:147–162
Andrews PW, Meyer LJ, Bednarz KL, Harris H (1984c) Two monoclonal antibodies recognizing determinants on human embryonal carcinoma cells react specifically with the liver isozyme of human alkaline phosphatase. Hybridoma 3:33–39
Andrews PW, Damjanov I, Simon D, Dignazio M (1985) A pluripotent human stem-cell clone isolated from the TERA-2 teratocarcinoma line lacks antigens SSEA-3 and SSEA-4 in vitro, but expresses these antigens when grown as a xenograft tumor. Differentiation 29:127–135
Andrews PW, Gönczöl E, Plotkin SA, Dignazio M, Oosterhuis JW (1986) Differentiation of TERA-2 human embryonal carcinoma cells into neurons and HCMV permissive cells: induction by agents other than retinoic acid. Differentiation 31:119–126
Andrews PW, Trinchieri G, Perussla B, Baglioni C (1987) Induction of class I major histocompatibility complex antigens in human teratocarcinoma cells by interferon without induction of differentiation, growth inhibition or resistance to viral infection. Cancer Res 47:740–746

Andrews PW, Gönczöl E, Fenderson B, Holmes EH, O'Malley G, Hakomori S-I, Plotkin SA (1989) Human cytomegalovirus induces stage-specific embryonic antigen-1 in differentiating human teratocarcinoma cells and fibroblasts. J Exp Med 169:1347–1359

Andrews PW, Nudelman E, Hakomori S, Fenderson BA (1990) Different patterns of glycolipid antigens are expressed following differentiation of TERA-2 human embryonal carcinoma cells induced by retinoic acid, HMBA or BUdR. Differentiation 43:131–138

Benham FJ, Andrews PW, Bronson DL, Knowles BB, Harris H (1981) Alkaline phosphatase isozymes as possible markers of differentiation in human teratocarcinoma cell lines. Dev Biol 88:279–287

Bennett D, Boyse EA, Old LJ (1972) Cell surface immunogenetics in the study of morphogenesis. In: Silvestri LG (ed) Cell interactions: third Lepetit symposium North Holland, Amsterdam, pp 247–263

Bird JM, Kimber SJ (1984) Oligosaccharides containing fucose linked α(1→3) and α(1→4) to *N*-acetylglucosamine cause decompaction of mouse morulae. Dev Biol 104:449–460

Bono R, Cartron JP, Mûlet C, Avner P, Fellous M (1981) Selective expression of blood group antigens on human teratocarcinoma cell lines. Blood Transfus Immunohaematol 24:97–107

Boyse EA, Old LJ (1978) The immunogenetics of differentiation in the mouse. Harvey Lect 71:23–53

Bremer EG, Levery SB, Sonnino S, Ghidoni R, Canevari S, Kannagi R, Hakomori S (1984) Characterization of a glycosphingolipid antigen defined by the monoclonal antibody MBrl expressed in normal and neoplastic epithelial cells of human mammary gland. J Biol Chem 259:14773–14777

Carlin CR, Andrews PW (1985) Human embryonal carcinoma cells express low levels of functional receptor for epidermal growth factor. Exp Cell Res 159:17–26

Chen C, Fenderson BA, Andrews PW, Hakomori S-I (1989) Glycolipid-glycosyltransferases in human embryonal carcinoma cells during retinoic acid-induced differentiation. Biochemistry 28:2229–2238

Clausen H, Levery SB, Nudelman E, Baldwin M, Hakomori S (1986) Further characterization of type 2 and type 3 chain blood group A glycosphingolipids from human erythrocyte membranes. Biochemistry 25:7075–7085

Cotte CA, Easty GC, Neville AM (1981) Establishment and properties of human germ cell tumors in tissue culture. Cancer Res 41:1422–1427

Damjanov I, Andrews PW (1983) Ultrastructural differentiation of a clonal human embryonal carcinoma cell line in vitro. Cancer Res 43:2190–2198

Dodd J, Jessel TM (1985) Lactoseries carbohydrate specify subsets of dorsal root ganglion neurons projecting to the superficial dorsal horn of the rat spinal cord. J Neurosci 5:3278–3294

Dodd J, Jessel TM (1986) Cell surface glycoconjugates and carbohydrate-binding proteins: possible recognition signals in sensory neuron development. J Exp Biol 124:225–238

Eisenbarth GS, Walsh FS, Nirenberg M (1979) Monoclonal antibody to a plasma membrane antigen of neurons. Proc Natl Acad Sci USA 76:4913–4917

Engström W, Rees AR, Heath JK (1985) Proliferation of a human embryonal carcinoma-derived cell line in serum free medium: interrelationship between growth factor requirements and membrane receptor expression. J Cell Sci 73:361–373

Fenderson BA, Zehavi U, Hakomori S (1984) A multivalent lacto-*N*-fucopentaose III-lysyllysine conjugate decompacts preimplantation-stage mouse embryos, while the free oligosaccharide is ineffective. J Exp Med 160:1591–1596

Fenderson BA, Andrews PW, Nudelman E, Clausen H, Hakomori S-I (1987) Glycolipid core structure switching from globo- to lacto- and ganglio-series during retinoic acid-induced differentiation of TERA-2-derived human embryonal carcinoma cells. Dev Biol 122:21–34

Gönczöl E, Andrews PW Plotkin SA (1984) Cytomegalovirus replicates in differentiated but not undifferentiated human embryonal carcinoma cells. Science 224: 159–161

Hakomori S (1981) Blood group ABH and Ii antigens of human erythrocytes: chemistry, polymorphism, and their developmental change. Semin Hematol 18: 39–62

Hass PE, Wada HG, Herman MM, Sussman HH (1979) Alkaline phosphatase of mouse teratocarcinoma stem cells: immunochemical and structural evidence for its identity as a somatic gene product. Proc Natl Acad Sci USA 76:1164–1168

Hirka G, Kawashima H, Prakash K, Plotkin S, Andrews PW, Gönczöl E (1991) Differentiation of human embryonal carcinoma cells induces HIV permissiveness, which is stimulated by HCMV co-infection. J. Virology 65:2732–2735

Hustin J, Collette J, Franchimont P (1987) Immunohistochemical demonstration of placental alkaline phosphase in various states of testicular development and in germ cell tumors. Int J Androl 10:29–35

Kannagi R, Nudelman E, Levery SB, Hakomori S (1982) A series of human erythrocyte glycosphingolipids reacting to the monoclonal antibody directed to a developmentally regulated antigen, SSEA-1. J Biol Chem 257:14865–14874

Kannagi R, Cochran NA, Ishigami F, Hakomori S-I, Andrews PW, Knowles BB, Solter D (1983a) Stage-specific embryonic antigens (SSEA-3 and -4) are epitopes of a unique globo-series ganglioside isolated from human teratocarcinoma cells. EMBO J 2:2355–2361

Kannagi R, Levery SB, Ishigami F, Hakomori S, Shevinsky LH, Knowles BB, Solter D (1983b) New globoseries glycosphingolipids in human teratocarcinoma reactive with the monoclonal antibody directed to a developmentally regulated antigen, stage-specific embryonic antigen 3. J Biol Chem 258:8934–8942

Lafemina R, Hayward GS (1986) Constitutive and retinoic acid inducible expression of cytomegalovirus immediate early genes in human teratocarcinoma cells. J Virol 58:434–440

Lee VM -Y, Andrews PW (1985) Differentiation of NTERA-2 clonal human embryonal carcinoma cells into neurons involves the induction of all three neurofilament proteins. J Neurosci 6:514–521

Levine P (1977) Comments on hemolytic disease of newborn due to anti-$PP_1P^k$ (Anti-$Tj^a$). Transfusion 17:573–578

Lopez M, Cartron J, Cartron JP, Mariotti M, Bony V, Salmon C, Levene C (1983) Cytotoxicity of anti-$PP_1P^k$ antibodies and possible relationships with early abortions of p mothers. Clin Immunol Immunopathol 28:296–303

Maio J, Brown FL (1988) Regulation of expression driven by human immunodeficiency virus type 1 and human T-cell leukemia virus type 1 long terminal repeats in pluripotential human embryonic cells. J Virol 62:1398–1407

Marton RW, Wiznia AA, Hutcheon RG, Rubinstein A (1986) Human T-cell lymphotrophic virus type III (HTLV-III) embryopathy: a new dysmorphic syndrome associated with intrauterine HTLV-III infection. Am J Dis Child 140: 638–640

Matthaei K, Andrews PW, Bronson DL (1983) Retinoic acid fails to induce differentiation in human teratocarcinoma cell lines that express high levels of cellular receptor protein. Exp Cell Res 143:471–474

Mavilio F, Simeone A, Boncinelli E, Andrews PW (1988) Activation of four homeobox gene clusters in human embryonal carcinoma cells induced to differentiate by retinoic acid. Differentiation 37:73–79

Muramatsu T, Gachelin G, Jacob F (1979) Characterization of glycopeptides from membranes of F9 embryonal carcinoma cells. Biochim Biophys Acta 587:392–406

Muramatsu H, Muramatsu T, Avner P (1982) Biochemical properties of high molecular weight glycopeptides released from the cell surface of human teratocarcinoma cells. Cancer Res 42:1749–1752

Naiki M, Marcus DM (1974) Human erythrocyte P and $P^k$ blood group antigens: identification as glycosphingolipids. Biochem Biophys Res Commun 60:1105–1111

Nelson JA, Groudine M (1986) Transcriptional regulation of the human cytomegalovirus major immediate early gene is associated with induction of DNase I-hypersensitive sites. Mol Cell Biol 6:425–461

Paiva J, Damjanov I, Lange PH, Harris H (1983) Immunohistochemical localization of placental-like alkaline phosphatase in testis and germ-cell tumors using monoclonal antibodies. Am J Pathol 111:156–165

Pera MF, Blasco-Lafita MJ, Cooper S, Mason M, Mills J, Monaghan P (1988) Analysis of cell-differentiation lineage in human teratomas using new monoclonal antibodies to cytostructural antigens of embryonal carcinoma cells. Differentiation 39:139–149

Pera MF, Cooper S, Mills J, Parrington JM (1989) Isolation and characterization of a multipotent clone of human embryonal carcinoma cells. Differentiation 42:10–23

Race RR, Sanger R, (1975) Blood groups in man 6th edn. Blackwell, Oxford

Rendt J, Erulkar S, Andrews PW (1989) Presumptive neurons derived by differentiation of a human embryonal carcinoma cell line exhibit tetrodotoxin-sensitive sodium currents and the capacity for regenerative responses. Exp Cell Res 180: 580–584

Rettig WJ, Cordon-Cardo C, NG JSC, Oettgen HF, Old LJ, Lloyd KO (1985) High molecular weight glycoproteins of human teratocarcinoma defined by monoclonal antibodies to carbohydrate determinants. Cancer Res 45:815–821

Sekiya S, Kawata M, Iwasawa H, Inaba N, Sugita M, Suzuki N, Motoyama T, Yamamoto T, Takamizawa H (1985) Characterization of human embryonal carcinoma cell lines derived from testicular germ-cell tumors. Differentiation 29:259–267

Shevinsky LH, Knowles BB, Damjanov I, Solter D (1982) Monoclonal antibody to murine embryos defines a stage-specific embryonic antigen expressed on mouse embryos and human teratocarcinoma cells. Cell 30:697–705

Soderstrom T, Enskogh A, Samuelsson BE, Cedergren B (1984) IgG3-restriction of anti-P and anti-$P^k$ antibodies may explain the early abortions in mothers of the p blood group (Abstr) 18th Congress of the International Society of Blood Transfusion, Munich

Solter D, Knowles BB (1978) Monoclonal antibody defining a stage-specific mouse embryonic antigen (SSEA-1). Proc Natl Acad Sci USA 75:5565–5569

Stagno S, Pass RF, Dworsky ME Alford CA (1983) Congenital and perinatal cytomegalovirus infections. Semin Perinatol 7:31–42

Strickland S, Mahdavi V (1978) The induction of differentiation in teratocarcinoma stem cells by retinoic acid. Cell 15:393–403

Suurmeijer AJH, Oosterhuis JW, Marrink J, de Bruijn HWA, Schraffordt Koops H, Sleijfer DT, Fleuren GJ (1983) Non-seminomatous germ cell tumors of testis. Analysis of AFP and HCG production by primary tumor and retroperitoneal lymph node metastases after PVB combination chemotherapy. Oncodev Bio Med 4:289–308

Teshima S, Shimosato Y, Hirohashi S, Tome Y, hayashi I, Kanazawa H, Kakizoe T (1988) Four new human germ cell tumor cell lines. Lab Invest 59:328–336

Thaller C, Eichele G (1987) Identification and spatial distribution of retinoids in the developing chick limb bud. Nature 327:625–628

Thompson S, Stern PL, Webb M, Walsh FS, Engstrom W, Evans EP, Shi WK, Hopkins B, Graham CF (1984) Cloned human teratoma cells differentiate into neuron-like cells and other cell types in retinoic acid. J Cell Sci 72:37–64

Thurin J, Herlyn M, Hindsgaul O, Stromberg N, Karlsson K, Elder D, Steplewski Z, Koprowski H (1985) Proton NMR and fast atom bombardment mass spectrometry analysis of the melanoma-associated ganglioside 9-*0*-acetyl-$GD_3$. J Biol Chem 260:14556–14563

Tippett P, Sanger R, Race RR, Swanson J, Busch S (1965) An agglutinin associated with the P and the ABO blood group systems. Vox Sang 10:269–280

Tippett P, Andrews PW, Knowles BB, Solter D, Goodfellow PN (1986) Red cell antigens P (globoside) and Luke: identification by monoclonal antibodies defining the murine stage-specific embryonic antigens -3 and -4 (SSEA-3 and -4). Vox Sang 51:53–56

Tsutsui Y, Nogami T, Sano M, Kashiwai A, Kato K (1987) Induction of S-100b (ββ) protein in human teratocarcinoma cells. Cell Differ. 21:137–145

Uchida T, Shimoda T, Miyata H, Shikata T, Iino S, Suzuki H, Oda T, Hirano K, Sugiara M (1981) Immunoperoxidase study of alkaline phosphatase in testicular tumors. Cancer 48:1455–1462

Wahren B, Holmgren PA, Stigbrand T (1979) Placental alkaline phosphatase, alpha fetoprotein and carcinoembryonic antigen in testicular tumor: tissue typing by means of cytological smears. Int J Cancer 24:749–753

Wan Y. -JY, Orrison BM, Lieberman R, Lazarovici P, Ozato K (1987) Induction of major histocompatibility class I antigens by interferons in undifferentiated F9 cells. J Cell Physiol 130:276–283

Weima SM, van Rooijen MA, Mummery CL, Feijen A, Kruijer W, de Laat SW, van Zoelen EJJ (1988) Differentially regulated production of platelet-derived growth factor and of transforming growth factor beta by a human teratocarcinoma cell line. Differentiation 38:203–210

Willison KR, Karol RA, Suzuki A, Kundu SK, Marcus DM (1982) Neutral glycolipid antigens as developmental markers of mouse teratocarcinoma and early embryos: an immunologic and chemical analysis. J Immunol 129:603–609

Wood JN, Hovanessian AG (1979) Interferon enhances 2-5A synthetase in embryonal carcinoma cells. Nature 282:74–76

# *Lactate Dehydrogenase Isoenzyme 1 in Testicular Germ Cell Tumors*

F.E. von Eyben, O. Blaabjerg, P.H. Petersen, S. Mommsen, E.L. Madsen, F. Kirpekar, S.S.-L. Li, and K. Kristiansen

Department of Clinical Chemistry, Odense University Hospital, Odense, Denmark

## Introduction

In his thesis, Zondag (1964) described the lactate dehydrogenase (L-lactate: $NAD^+$-oxidoreductase EC 1.1.1.27, LDH) isoenzyme patterns of human malignant diseases. Among other findings, he drew attention to the characteristic LDH isoenzyme pattern in germ cell tumors, which showed a predominance of LDH isoenzyme 1 (LDH-1), being deviant from those of other cancers. This LDH-1 pattern in tumor tissue has been confirmed in later investigations (Murakami and Said 1984; Takeuchi et al. 1979). Similarly, the serum activity is frequently raised in patients with tumors (von Eyben 1983). The increased LDH-1 level in testicular tumors could be due to polyploidization involving the gene locus encoding LDH-B, as LDH-1 is a homotetrameric combination of this subunit. The *LDH-B* gene is localized on the short arm of chromosome 12 (12p12.1–12p12.2), and multiple copies of isochromosome i(12p) are often present in human testicular germ cell tumors (Atkins and Baker 1982). The raised tissue LDH-1 could also be related to an increase in the rate of transcription of the *LDH-B* gene, an increased stability of the LDH-B mRNA, an increase in the translation of the LDH-B mRNA, or a combination of all these factors (Fig. 1).

As an approach to elucidate the background for the LDH-1 isoenzyme pattern in human testicular germ cell tumors, we measured the changes in the levels of human LDH-B mRNA in tumor tissue and in normal testicular tissue. In addition we determined the levels of transcripts from the proto-oncogene c-Ki-*ras2*, which like the *LDH-B* gene is localized on the short arm of chromosome 12 (12p12.1) (McKusick 1986). We compared the LDH-1 activity of tumor tissue with that in normal testicular tissue, and measured S-LDH-1 in blood from the testicular vein and a peripheral vein at orchiectomy of the tumor patients. Finally, we compared the S-LDH-1 level in peripheral blood from tumor patients with that of controls.

Recent Results in Cancer Research, Vol. 123

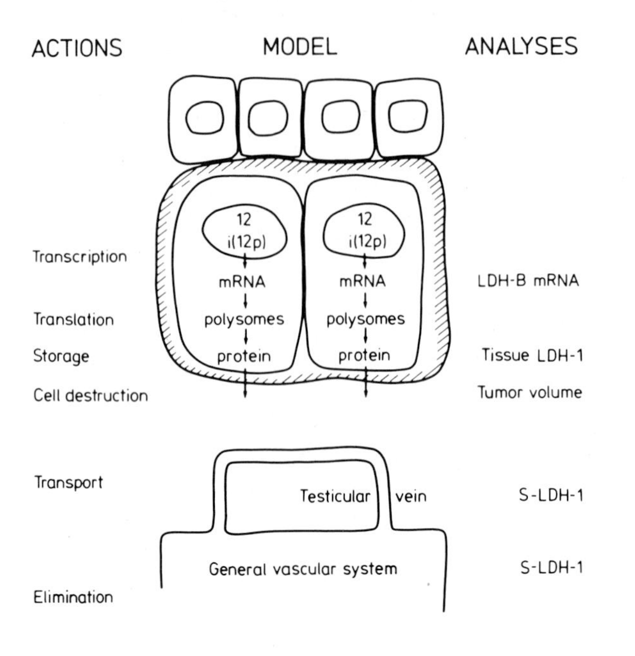


**Fig. 1.** Model and assays

## Materials and Methods

*Patients.* Eighteen men were included in the study. All were admitted between August 1988 and July 1989 to the Dept. of Urology, Odense University, due to suspected testicular germ cell malignancy. Seven had a testicular germ cell tumor and underwent orchiectomy. Their median age was 45 years (range, 33–49 years). Six had a seminoma and one a nonseminomatous tumor (embryonal carcinoma, teratoma, and choriocarcinoma). Five had a stage 1 tumor, one a stage 2 tumor, and one a stage 3 tumor. Eleven had no tumor on surgical exploration. Their median age was 36 years (range, 20–52 years). Three had epididymitis, two periorchitis, two hydrocele, one Leydig's cell hyperplasia, one testicular torsion, one an appendicular cyst, and one no abnormality of the testis.

*Tissue Samples from Testicular Tumor and Nonmalignant Testis.* We obtained tissue from the malignant part of the testis in all tumor patients as well as from the macroscopically normal part of the testis in four. However, microscopic examination of this part of the testis from two of the four patients revealed only tumor tissue. A testis with Leydig's cell hyperplasia orchiectomized from one of the controls also served to estimate the LDH-1 level of nonmalignant testicular tissue.

*RNA Isolation.* Tissue was ground under liquid nitrogen and RNA was extracted by the hot phenol-guanidinium thiocyanate procedure (Maniatis et al. 1982). RNA was dissolved in water and the concentration was determined spectrophotometrically.

*RNA Slot Blotting.* RNA was denatured by treatment with glyoxal (McMaster and Carmichael 1977). From each sample 0.2-μg, 1-μg, and 5-μg portions of the denatured total RNA were deposited on Zeta-Probe (Bio-Rad) membranes using a Bio-Rad Bio-Dot SF blotting apparatus.

*Hybridization of Filters.* Gel-purified fragments were used as probes and $^{32}P$-labeled by the random primer method (Feinberg and Vogelstein 1984). The *LDH-B* probe was a 1.3-kb *Eco*RI cDNA fragment (Sakai et al. 1987). The probe for c-Ki-*ras2* transcripts was a 1.8-kb *Xba*I-*Eco*RI fragment containing Ki-*ras* exon 4B isolated from pCD1234B (McCoy and Weinberg 1986), and the $\beta_2$-micro-globulin probe was a 0.54-kb *Pst*I cDNA fragment (Suggs et al. 1981). Hybridizations were performed in buffer containing 40% deionized formamide and 10% dextran sulfate (Maniatis et al. 1982). The filters were autoradiographed at −70°C with one intensifying screen. The autoradiographs were analyzed by scanning densitometry using a Shimadzu CS-9000 scanning densitometer.

*Estimation of Levels of LDH-B and c-Ki-*ras2 *mRNAs.* The amounts of mRNA for LDH-B and c-Ki-*ras2* were estimated by the intensity of the autoradiographic hybridization signal relative to that of $\beta_2$-microglobulin mRNA. The results were analyzed within the range of values that had been found to reveal a log-log linear relationship between the amount of RNA and the intensity of the signal.

*Estimation of Total Tumor mRNA for LDH-B.* We used the product of the relative amount of mRNA and tumor volume, calculated according to the formula for a revolving ellipsoid (von Eyben et al. 1988), as an estimate of the total tumor mRNA for LDN-B.

*Tissue LDH-1 Determinations.* The tissue specimens from the testis were frozen and stored at −80°C immediately after the resection of the testis. We homogenized the tissue samples in a phosphate buffer and sonicated the homogenate. The volume was adjusted to 10 or 40 ml. After short ultra-centrifugation, we determined the LDH-1 in the supernatant using our serum assay (von Eyben et al. 1988).

*Estimation of Total Tumor LDH-1.* We used the product of the tumor volume and the activity of LDH-1 in tumor tissue (U/l) as an estimate of the total isoenzyme activity in the tumor lesion.

*S-LDH-1 Determinations.* Blood samples, S-LDH-1 determinations, correction for hemolysis, and calculations of the baseline level of S-LDH-1 in the tumor patients were obtained and performed as described previously (von Eyben et al. 1988). The upper limit of the reference values was 109 U/l.

*Estimation of Tumor Impact on S-LDH-1 in Testicular Vein and Peripheral Arm Vein Blood.* We estimated the release of LDH-1 from the testicular tumor as the difference between the preoperative S-LDH-1 in testicular vein and peripheral arm vein blood. We used the difference between the preoperative S-LDH-1 in peripheral arm vein blood and the S-LDH-1 after normalization following orchiectomy as an estimate of the contribution from the tumor to S-LDH-1 in the general vascular system.

## Results

*mRNA of LDH-B and c-Ki-*ras *in Normal and Malignant Testicular Tissue.* In macroscopically malignant testicular tissue, the median value for the level of LDH-B mRNA relative to that in normal testicular tissue was 2.0 (range, 0.5–18.4). It was 1.9 and 4.2 in two microscopically normal testes. In macroscopically malignant testicular tissue, the median value for the level of c-Ki-*ras2* mRNA relative to that in normal testis was 1.1 (range, 0.6–2.5). It was 0.5 and 1.9 in two microscopically normal testes (Table 1).

*LDH-1 in Testicular Tumor Tissue and Normal Testicular Tissue.* The LDH-1 activity of the tumors of the seven patients is shown in Table 2. The median tumor tissue LDH-1 was 64 U/g (range, 17–286 U/g). The corresponding activity was 5.2 and 106 U/g in testicular tissue from two patients, which was judged macroscopically to be normal but was found microscopically to be malignant. It was median 18 U/g in normal testicular tissue.

*Testicular Vein Blood S-LDH-1.* S-LDH-1 level in testicular vein blood was raised (median, 340 U/l; range, 125–889 U/l) in connection with orchiectomy in five patients with tumor (Table 2).

*S-LDH-1 in Peripheral Arm Vein Blood.* S-LDH-1 had a median value of 80 U/l (range, 55–286 U/l) in the control patients (Table 2) and a median value of 159 U/l (range, 68–728 U/l) in the tumor patients. S-LDH-1 levels in the two groups differed significantly ($p = 0.023$, Mann Whitney U test, one-tailed). Eight of 11 controls had a S-LDH-1 level below 109 U/l and three had higher levels. In contrast, six of the seven tumor patients had raised S-LDH-1 levels and only one an activity within the reference interval (Table 2, Fisher's exact text, $p = 0.025$ one-tailed test)

*Relationship Between Total Tumor mRNA for LDH-B and Raise in S-LDH-1.* Total tumor tissue mRNA for LDH-B correlated with the raise in testicular vein S-LDH-1 (Spearman's rank correlation coefficient $r = 0.6$, $p = 0.16$,

**Table 1.** Tumor size, tumor stage, and mRNA levels

| Patient No. | Tumor size (ml) | Stage | Histology | Relative amount of mRNA LDH-B in macroscopically | | c-Ki-*ras2* in macroscopically | |
|---|---|---|---|---|---|---|---|
| | | | | normal testis | malignant testis | normal testis | malignant testis |
| 1 | – | – | – | 1 | – | 1 | |
| 4 | 22 | 1 | S | 1.9 | 14.8 | 1.9 | 2.5 |
| 7 | 12 | 1 | S | – | 1.6 | – | 1.9 |
| 8 | 264 | 1 | S | – | 0.5 | – | 0.6 |
| 9 | 126 | 2 | S | 2.8[a] | 1.2 | 0.8[a] | 0.6 |
| 12 | 79 | 3 | NS | 4.2 | 4.2 | 0.5 | 1.1 |
| 16 | 83 | 1 | S | 2.4[a] | 18.4 | 1.3[a] | 1.4 |
| 17 | 31 | 1 | S | – | 2.0 | – | 1.1 |

[a] Histologically malignant tissue only.
The levels of mRNA for LDH-B and for c-ki-*ras2* in normal and malignant testicular tissue are given relative to the level of mRNA for $\beta_2$-microglobulin with the nonmalignant testis from patient 1 as control. The control values are the average of two independent determinations on two tissue samples. *S* denotes seminoma; *NS*, a nonseminomatous tumor.

**Table 2.** Levels of LDH-1 in malignant and nonmalignant testicular tissue from eight patients

| Patient No. | LDH-1 (U/g) in macroscopically | | S-LDH-1 (U/l) in testicular vein | S-LDH-1 (U/l) in arm vein |
|---|---|---|---|---|
| | normal testis | malignant testis | | |
| 1 | 8.1 | – | – | 55 |
| 4 | 18 | 64 | 235 | 159 |
| 7 | – | 66 | – | 68 |
| 8 | – | 34 | 340 | 157 |
| 9 | 106[a] | 286 | 889 | 728 |
| 12 | 19 | 17 | – | 619 |
| 16 | 5.2[a] | 43 | 657 | 222 |
| 17 | – | 117 | 125 | 130 |

[a] Histologically malignant tissue only.

two-tailed test, Fig. 2). It also correlated with the raise in peripheral arm vein S-LDH-1 ($r = 0.8$, $p = 0.16$, two-tailed test).

*Relationship Between the Raised S-LDH-1 Level in the Testicular Vein and the Peripheral Arm Vein.* The rise in testicular vein blood and peripheral arm vein S-LDH-1 levels were significantly correlated (Spearman's rank correlation coefficient $r = 0.80$, $p = 0.16$, two-tailed test, Fig. 3).

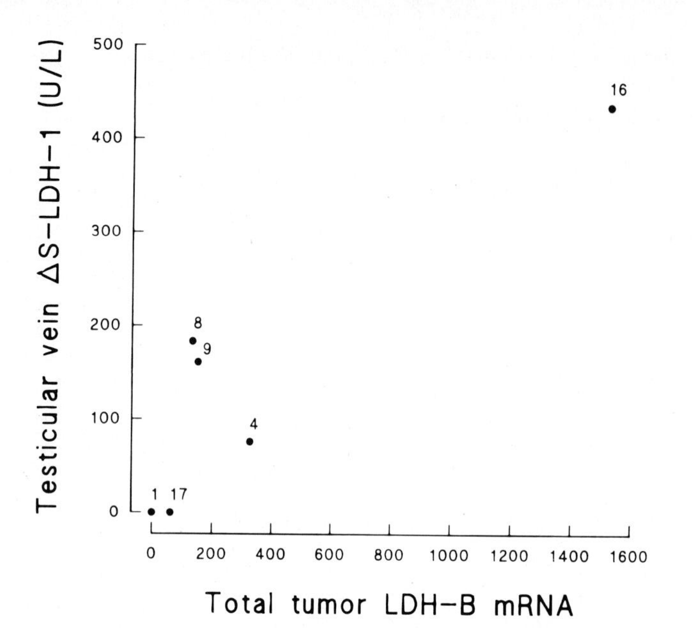


**Fig. 2.** Relationship between the total tumor mRNA for LDH-B and the raise in testicular vein S-LDH-1. The *abscissa* shows the total tumor mRNA as the product of the relative amount of mRNA and the volume of the primary tumor. The ordinate shows the release to the testicular vein given as the difference between the activity in the testicular vein at orchiectomy and that in the peripheral arm vein

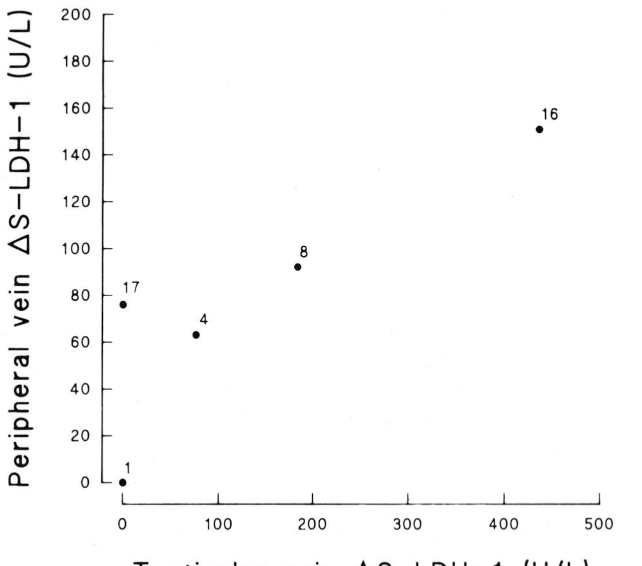

**Fig. 3.** Output of LDH-1 at orchiectomy in five patients. The *abscissa* shows the release to the testicular vein given as mentioned in Fig. 2. The *ordinate* shows the release to the peripheral vein given as the difference between the orchiectomy S-LDH-1 level and the normalized postoperative S-LDH-1 level. 1 denotes a patient without malignancy. All other patients had tumor stage 1

*Relationship Between Stage and S-LDH-1.* Arm vein S-LDH-1 was significantly higher in stage 2 and 3 patients than in stage 1 patients (Mann Whitney U test, $p = 0.047$, one-tailed test).

## Discussion

The present study provides additional evidence that the predominance of LDH-1 in human testicular germ cell tumor is a cancer-related nonrandom isoenzymatic abnormality. Most testicular germ cell tumor tissue had a relatively high level of mRNA for LDH-B and a high level of LDH-1. We found no significant increase in the level of c-Ki-*ras2* mRNA.

In the majority of the tumor patients, testicular vein S-LDH-1 activity was higher than in the peripheral vein. The rise in S-LDH-1 in the testicular vein correlated significantly with the rise in S-LDH-1 in the peripheral vein at orchiectomy. Peripheral arm vein blood S-LDH-1 level was higher in the tumor patients than in the controls.

The absence of an increase in the level of c-Ki-*ras2* mRNA was surprising given the frequent finding of i(12p) in testicular germ cell tumors (Atkins and Baker 1982). The minor fluctuations of c-Ki-*ras2* mRNA, contrasting with much greater fluctuations in the level of LDH-B mRNA, indicates that the increased abundance of the latter transcript is not solely caused by a gene dosage mechanism due to the possible presence of i(12p). Thus, the marked increase in the abundance of LDH-B mRNA must be based on a pronounced derepression of the *LDH-B* gene, a marked increase in the stability of the LDH-B mRNA, or both.

Previous studies of human disorders such as myocardial infarction and hemolytic anemia (Herscher et al. 1984) as well as experimental cell culture investigations (Kristensen and Hørder 1988) indicate that LDH-1 is released from living cells as they reach the point of irreversible damage. It is therefore likely that the release of LDH-1 from testicular germ cell tumors also in based on irreversible cellular changes, such as cell loss.

The relation between total mRNA for LDH-B and the raise in testicular vein S-LDH-1 suggests that the release of LDH-1 from the tumor reflects the production of LDH-1. Similarly, the relationship between total tumor mRNA for LDH-1 and the raise in peripheral arm vein S-LDH-1 links the production of LDH-1 in the tumor to the peripheral level of S-LDH-1.

In conclusion, the present study strongly points to the testicular germ cell tumor as the source of the raised S-LDH-1. The tumors appear most often to have an increased level of mRNA for LDH-1, reflected in an increased production and release of LDH-1. The high level of mRNA might be due to an increased number of the short arm of chromosome 12 in most tumors.

## *References*


Atkins NB, Baker MC (1982) Specific chromosome change i(12p) in testicular tumors? Lancet 2:1349

Feinberg AP, Vogelstein B (1984) Addendum. A technique for radiolabelling DNA restriction endonuclease fragments to high specific activity. Anal Biochem 137: 266–267

Herscher LL, Siegel RJ, Said JW, Edwards GM, Moran MM, Fishbein MC (1984) Distribution of LDH-1 in normal, ischemic and necrotic myocardium. Am J Clin Pathol 81:198–203

Kristensen SR, Hørder M (1988) Release of enzymes from quiescent fibroblasts during ATP depletion. Enzyme 39:205–212

Maniatis T, Fritsch EF, Sambrook J (1982) Molecular cloning: a laboratory manual. Cold Spring Harbor Labratory, Cold Spring Harbor

McCoy MS, Weinberg RA (1986) A human Ki-*ras* oncogene encodes two transforming p21 proteins. Mol Cell Biol 6:1326–1328

McKusick VA (1986) The human gene map 15 April 1986. Clin Genet 29:545–588

McMaster GK, Carmichael GG (1977) Analysis of single and double stranded nucleic acids on polyacrylamide and agarose gels by using glyoxal and acridine orange. Proc Natl Acad Sci USA 74:4835–4838

Murakami SS, Said JW (1984) Immunohistochemical localization of lactate dehydrogenase isoenzyme 1 in germ cell tumors of the testis. Am J Clin Pathol 81:293–296

Sakai I, Sharief FS, Pan Y-CE, Li SS-L (1987) The cDNA and protein sequences of human lactate dehydrogenase B. Biochem J 248:933–936

Suggs SV, Wallace RB, Hirose T, Kawashima EH, Itakura K (1981) Use of synthetic oligonucleotides as hybridization probes: isolation of cloned cDNA sequences for human $\beta_2$-microglobulin. Proc Natl Acad Sci USA 78:6613–6617

Takeuchi T, Nakayasu M, Hirohashi S, Kameya T, Kaneko M, Yokomori K, Tsuchida Y (1979) Human endodermal sinus tumour in nude mice and its markers for diagnosis and management. J Clin Pathol 2:693–699

von Eyben FE (1983) Lactate dehydrogenase and its isoenzymes in testicular germ cell tumors: an overview. Oncodev Biol Med 3:395–414

von Eyben FE, Blaabjerg O, Hyltoft Petersen P, Hørder M, Nielsen HV, Kruse-Andersen S, Parlev E (1988) Serum lactate dehydrogenase isoenzyme 1 as a tumor marker of testicular germ cell tumor. J Urol 140:986–990

Zondag HA (1964) Bepaling en diagnostische betekenis van de melkzuurdehydrogenase isoenzymen. Thesis, University of Groningen

# *Expression of Developmentally Regulated Genes in Embryonal Carcinoma Cells*

P. Dráber

Institute of Molecular Genetics, Czechoslovak Academy of Sciences, Vídeňská 1083, 142 20 Prague 4, Czechoslovakia

## Introduction

Human and mouse embryonal carcinoma (EC) cell lines are widely used for analysis of early mammalian embryogenesis and germ cell neoplasia (Martin 1980; Andrews et al. 1983b). Using monoclonal antibodies raised against EC cells it has been shown that these cells express several cell surface epitopes shared with early embryonic cells. The first stem cell epitope identified by the hybridoma technology was the epitope of an antigen named stage-specific embryonic antigen 1 (SSEA-1; Solter and Knowles 1978). This epitope was found on mouse preimplantation stage embryos from the morula stage on and also in the brain and kidney and several other organs of adult mice. The epitope was expressed in all mouse EC cell lines and it disappeared from most of these cells during their differentiation in vitro. In contrast to mouse EC cells, the human EC cells expressed low amounts of SSEA-1, and during differentiation the amount of SSEA-1 was significantly increased (Andrews et al. 1983b). A number of other developmentally regulated epitopes have been discovered on the surface of mouse and human EC cells. Biochemical analysis of these epitopes revealed that they are carbohydrate portions of membrane glycoconjugates (for review, see Feizi 1985; Hakomori 1985; Muramatsu 1988).

Although the developmentally regulated carbohydrate epitopes are extremely useful as markers for analysis of embryonic stem cell differentiation and germ cell neoplasia, they are not amenable to direct genetic analysis. So far, molecular analysis of genes for developmentally regulated surface antigens has been limited to those genes which are poorly expressed in stem cells but efficiently transcribed after their differentiation. These include *H-2* (Croce et al. 1981), plasminogen activator (Strickland et al. 1980; Rickles et al. 1989), collagen type IV and laminin (Hogan 1980; Strickland et al. 1980), and α-fetoprotein (Young and Tilghman 1984). Molecular analysis of developmentally regulated genes for stem cell surface antigens requires production of antibodies recognizing protein epitopes of stem cell antigens and the use of these antibodies for isolation of corresponding genes from expression cDNA libraries.

In this laboratory we have recently described two surface epitopes which (a) are expressed in mouse EC cells, (b) disappear during differentiation of these cells in vitro, and (c) are carried by the protein moiety of surface glycoconjugates.

## Surface Antigens of Embryonal Carcinoma Cells

### *TEC-4 Antigen*

The first antigen, named TEC-4, has been identified by a monoclonal antibody prepared by a newly devised strategy, based on the facts that developmentally regulated carbohydrate epitopes carried by embryoglycan are extremely immunogenic and that most of the antibodies in animals immunized with EC cells are directed against embryoglycan. To overcome this problem, we produced embryoglycan-defective mutant cells and used them for immunization. Cells defective in the expression of embryoglycan were prepared from mutagenized P19 EC cells by a single-step selection technique using monoclonal antibody TEC-01 (Dráber and Pokorná 1984) conjugated with the plant toxin ricin (Dráber and Vojtišková 1984). Of the three independently isolated mutant cell lines, one apparently lacked embryoglycan as revealed by glycopeptide analysis and radioantibody-binding assays (Dráber and Malý 1987). Electorphoretic analysis of cell extracts from embryoglycan-negative mutant cells (P19XT. 1.1) revealed that mutant cells do not possess Gal ($\beta$1->4)-[Fuc($\alpha$1->3)]G1cNAc, the epitope recognized by TEC-01 (anti-SSEA-1) antibody (Fig. 1).

Although the mutant cells lacked carbohydrate structures typical of early embryonic and EC cells, they still exhibited many properties of EC cells including morphology and ability to differentiate in vitro (Dráber and Malý 1987). Injection of the mutant cells into rats induced an antibody response. Antibody-producing cells were immortalized by fusion with mouse myeloma cells, and a TEC-04 antibody producing clone was selected (Dráber et al. 1989b). The target epitope was found to be expressed on mouse EC and embryonic stem cells, but not on differentiated cell lines, except for neuroblastoma C1300/E7. During differentiation induced by retinoic acid, the TEC-4 epitope disappeared before the onset of morphological differentiation. Aggregates of P19 cells, formed by plating the cells into tissue culture dishes containing 1.5% Bacto-agar in culture medium, were TEC-4$^{+}$, whereas aggregates formed in the presence of 0.5 $\mu M$ retinoic acid were TEC-4$^{-}$ (Fig. 2). Immunochemical analysis of immunoaffinity-isolated TEC-4 antigen revealed that TEC-4 epitope is associated with glycoproteins of apparent $M_r$ 120000 and 240000. The epitope was resistant to oxidation by sodium

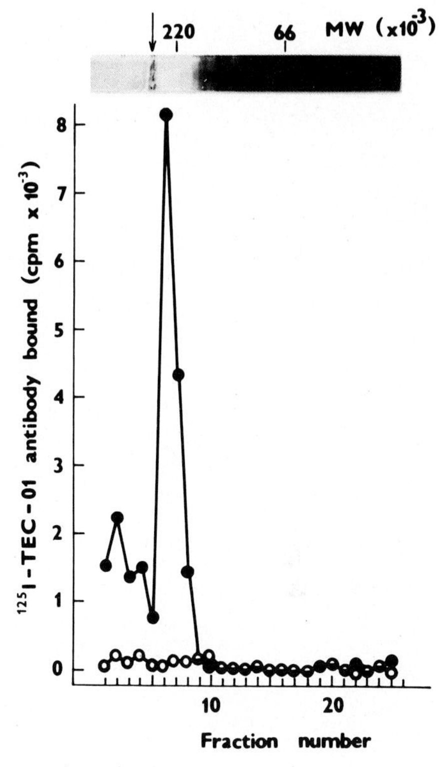


**Fig. 1.** Binding of TEC-01 antibody to electrophoretically separated NP-40 extracts from P19X1 cells (●) and TEC-01-ricin-resistant P19XT.1.1 cells (○). Cell extracts were subjected to SDS-PAGE, and strips of the gels were sliced into 2-mm segments. The antigens eluted from the slices were immobilized on plastic by evaporation and detected with $^{125}$I-labeled TEC-01 antibody as described (Dráber 1987). In the *upper part* the corresponding protein pattern revealed by staining with Coomassie brilliant blue is shown. The *arrow* indicates the stacking gel-separation gel interface. *220* and *66* indicate the migration peaks of ferritin ($M^r$, 220 000) and albumin ($M^r$ 66 000), respectively

periodate and to digestion by endoglycosidase F, but was sensitive to treatment with protein-denaturing agents and proteases. This suggested that TEC-4 epitope is located in the protein moiety of the molecule. Immunoaffinity-purified TEC-4 antigen is being used for production of polyclonal antisera with the aim of cloning the corresponding gene from the P19 expression cDNA library.

### *Thy-1 Antigen*

The second antigen has been discovered by screening of various monoclonal antibodies for binding to mouse EC cells. One monoclonal antibody 1aG4 has

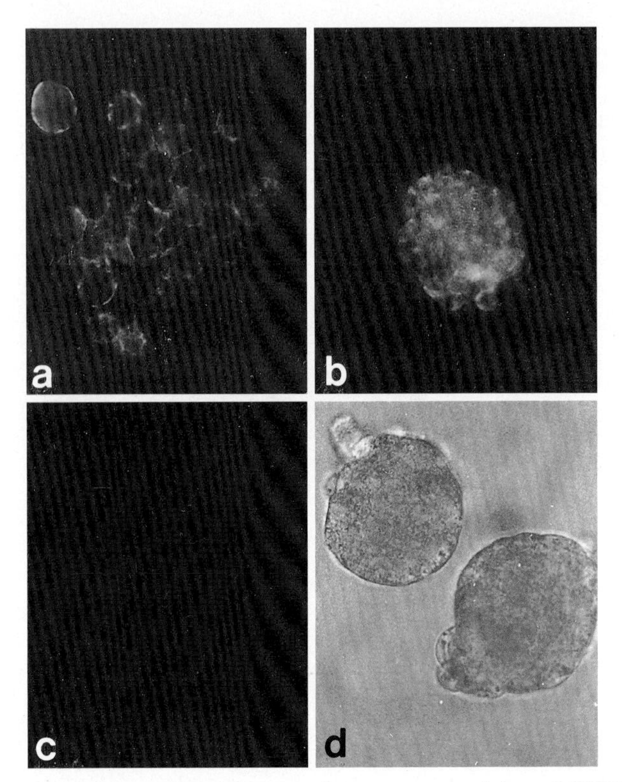


**Fig. 2a–d.** Indirect immunofluorescence staining of P19X1 cells with TEC-04 antibody. Cells cutlured for 3 days in tissue culture dish (**a**). Aggregate of cells after 3 days in culture on agar (**b**). Aggregate of cells after 3 days in culture on agar in the presence of 0.5 μ*M* retinoic acid (**c**, **d**). Photomicrographs were taken using ultraviolet (*UV*) illumination (**a–c**) or bright-field illumination (**d**)

been found to bind to P19 EC cells but did not bind to their differentiated derivatives and to a number of transformed cell lines (Dráber et al. 1989a). 1aG4 antibody recognizes mouse Thy-1.2 antigen (Dráber et al. 1980). Thy-1 is a glycoprotein with an apparent $M_r$ of 25000 which is bound to the cell surface through a phosphatidyl-inositol-containing membrane-binding domain (Williams and Gagnon 1982; Tse et al. 1985). In mice, Thy-1 exists in two allotypic forms, Thy-1.1 and Thy-1.2, which differ by a single amino acid substitution at residue 89 (arginine in Thy-1.1, glutamine in Thy-1.2) and can be distinguished by allele-specific monoclonal antibodies. Thy-1 antigen has been found in a number of mouse cell types including brain cells, thymus cells, splenic T cells, some peripheral nerves, smooth muscle and ganglion cells in stomach, kidney urothelium, connective tissue, and skin fibroblasts (Morris 1985; Rettig and Old 1989). Human Thy-1 antigen can be distinguished from mouse Thy-1.1 and Thy-1.2 antigen by monoclonal antibodies. Immunohistochemical analysis of human tissues revealed that human Thy-1 is also expressed in brain but not in kidney urothelium, thymus, and splenic T cells. On the other hand, several tissues which are Thy-1$^{-}$ in mice are Thy-1$^{+}$ in humans. These include kidney tubules and blood vessel endothelium. Human EC cells are Thy-1$^{+}$ (Andrews et al. 1983a), whereas mouse EC cells are Thy-1$^{+}$ (Dráber et al. 1989a) or Thy-1$^{-}$ (Stern et al. 1975). Interestingly, in the course of retinoic acid-induced differentiation, Thy-1 antigen disappears from the surface of P19 EC cells (Dráber et al. 1989a) but is not significantly modulated in human EC cells (Andrews et al. 1983a).

Because of these characteristic patterns of expression, Thy-1 is an interesting model for the investigation of regulatory mechanisms that determine its cell and tissue specificity. Furthermore, large genomic clones encompassing the human *Thy-1* and mouse *Thy-1.1* and *Thy-1.2* genes have been isolated and sequenced (Giguere et al. 1985; Seki et al. 1985; Ingraham et al. 1986).

To analyze the genetic elements that determine interspecies differences in Thy-1 antigen expression, we have introduced genomic DNA fragments containing the human *Thy-1* gene and mouse *Thy-1.1* gene into Thy-1.2 homozygous P19 EC cells and have examined changes in Thy-1 antigen expression during differentiation of the transfected cells (Malý and Dráber, in preparation). From a number of clones isolated we found several clones, which had morphology of EC cells and expressed either human Thy-1 antigen or mouse Thy-1.1 antigen, depending on the gene used for transfection. There is a rough correlation between number of gene copies and expression of Thy-1 antigen on cell surface as detected by direct radioantibody-binding assay. When the transfected cells were treated with retinoic acid, they differentiated into fibroblast-like cells. This differentiation was accompanied by a decrease in the expression of TEC-1 and TEC-4 epitopes (Fig. 3). In differentiated T2/1/B5 cells the endogenous *Thy-1.2* gene product exhibited more significant reduction in expression than the exogenous human *Thy-1* gene product. In T2/2/A4 cells with exogenous *Thy-1.1* gene, a similar reduction in the expression of both endogenous *Thy-1.2* and exogenous *Thy-1.1* genes was observed. These data are consistent with recent observations (Gordon et al. 1987) that mouse *Thy-1.1* gene in transgenic mice is expressed similarly to endogenous *Thy-1.2* gene, whereas human *Thy-1* gene is expressed according to the human pattern of *Thy-1* gene expression. However, analysis of a number of other P19-derived cell lines with exogenous mouse or human *Thy-1* genes revealed that the picture is more complex and that a number of factors such as initial stage of differentiation, number of exogenous *Thy-1* gene copies/genome, integration site, and others play a significant role (Malý and Dráber, in preparation).

The expression of polymorphic *Thy-1* gene product on the surface of P19 EC cells is also important for the understanding of cell reprogramming in cell hybrids. Hybrids between mouse lymphocytes and EC cells have properties of EC cells, i.e., the differentiated phenotype of the lymphoid parent is completely suppressed (Andrews and Goodfellow 1980; Rousset et al. 1980; Gmur et al. 1981; Forejt 1983). The lymphocyte genome in hybrid cells can be

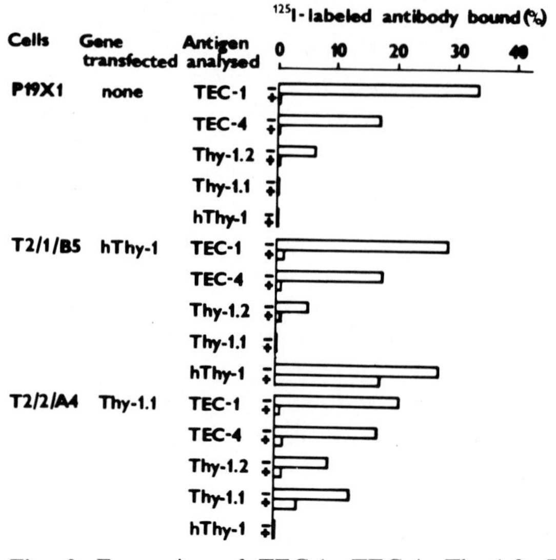


**Fig. 3.** Expression of TEC-1, TEC-4, Thy-1.2, Thy-1.1, and human (*h*) Thy-1 epitopes on the surface of P19X1 cells and cells derived from them by transfection with *hThy-1* or *Thy-1.1* gene. EcoRI fragments containing *Thy-1.1* or *hThy-1* gene were cotransfected with *pSTneoB* gene (Katoh et al. 1987). Colonies surviving in the antibiotic G418 were propagated and then tested in direct radioantibody-binding assay using $^{125}$I-labeled antibodies. (−) and (+) indicate the expression of the analyzed antigen in untreated cells and cells treated for 5 days with 0.5 μ*M* retinoic acid, respectively

extinguished or it can be reprogrammed to an embryo-specific stage. The latter possibility would imply that lymphocyte genome can be reprogrammed to the pluripotent stage and that lymphocyte differentiation is an irreversible process.

Because polymorphic SSEAs have not been identified, it has not been possible to elucidate the contribution of the lymphocyte genome in EC phenotype properties of the hybrids. Fusion of B lymphocytes from the Thy-1.1$^+$ mouse strain with P19 EC cells and analysis of the expression of Thy-1.1 and Thy-1.2 antigens in the hybrids may, at least in part, solve this problem.

## Conclusion

Mouse and human EC cells exhibit some unusual features which distinguish them from differentiated cells. They express a number of developmentally regulated stem cell antigens which disappear during their differentiation. Molecular genetic analysis of stem cell surface antigens requires availability of probes for the identification and analysis of these antigens and genes responsible for their expression. Monoclonal antibodies TEC-04 and 1aG4 recognizing protein epitopes of developmentally regulated surface antigens of mouse EC cells, TEC-4, and Thy-1, respectively, may represent the required probes.

### *References*

Andrews PW, Goodfellow PN (1980) Antigen expression by somatic cell hybrids of murine embryonal carcinoma cell with thymocytes and L cells. Somatic Cell Genet 6:271–284

Andrews PW, Goodfellow PN, Bronson DL (1983a) Cell-surface characteristics and other markers of differentiation of human teratocarcinoma cells in culture. Cold Spring Harbor Conf Cell Proliferation 10:579–590

Andrews PW, Goodfellow PN, Damjanov I (1983b) Human teratocarcinoma cells in culture. Cancer Surv 1:41–73

Croce CM, Linnenbach A, Huebner K, Parnes JR, Margulies DH, Appella E, Seidman JG (1981) Control of expression of histocompatibility antigens (H-2) and $\beta_2$-microglobulin in F9 teratocarcinoma stem cells. Proc Natl Acad Sci USA 78:5754–5758

Dráber P (1987) The epitope of mouse embryonic antigen(s) recognized by monoclonal antibody TEC-02 is a carbohydrate carried by high-molecular-weight glycoconjugates. Cell Differ 21:119–130

Dráber P, Malý P (1987) Mutants of embryonal carcinoma cells defective in the expression of embryoglycan. Proc Natl Acad Sci USA 84:5798–5802

Dráber P, Pokorná Z (1984) Differentiation antigens of mouse teratocarcinoma stem cells defined by monoclonal antibodies. Cell Differ 15:109–113

Dráber P, Vojtíšková M (1984) Developmentally regulated surface structures of teratocarcinoma stem cells studies by mutant cell lines. Cell Differ 15:249–253

Dráber P, Zikán J, Vojtíšková M (1980) Establishment and characterization of permanent murine hybridomas secreting monoclonal anti-Thy-1 antibodies. J Immunogenet 7:455–474

Dráber P, Malý P, Hausner P, Pokorná Z (1989a) The Thy-1 glycoprotein is expressed in mouse embryonal carcinoma cells P19. Int J Dev Biol 33:369–378

Dráber P, Nosek J Pokorná Z (1989b) Unusual stage-specific embryonic antigen (TEC-4) defined by a monoclonal antibody to embryonal carcinoma cells defective in the expression of embryoglycan. Proc Natl Acad Sci USA 86:9337–9341

Feizi T (1985) Demonstration by monoclonal antibodies that carbohydrate structures of glycoproteins and glycolipids are onco-developmental antigens. Nature 314: 53–57

Forejt J (1983) Expression of *T-t* and *H-2* genes in teratocarcinoma cells. Cold Spring Harbor Conf Cell Proliferation 10:541–552

Giguere V, Isobe K, Grosveld F (1985) Structure of murine *Thy-1* gene. EMBO J 4:2017–2024

Gmur R, Knowles BB, Solter D (1981) Regulation of phenotype in somatic cell hybrids derived by fusion of teratocarcinoma cell lines with normal or tumor-derived mouse cells. Dev Biol 81:245–254

Gordon JW, Chesa PG, Nishimura H, Rettig WJ, Maccari JE, Endo T, Seravalli E, Seki T, Silver J (1987) Regulation of *Thy-1* gene expression in transgenic mice. Cell 50:445–452

Hakomori S (1985) Aberrant glycosylation in cancer cell membranes as focused on glycolipids: overview and perspectives. Cancer Res 45:2405–2414

Hogan BLM (1980) High molecular weight extracellullar proteins synthesized by endoderm cells derived from mouse teratocarcinoma cells and normal extra-embryonic membranes. Dev Biol 76:275–285
Ingraham HA, Lawless GM, Evans GA (1986) The mouse *Thy-1.2* glycoprotein gene: complete sequence and identification of an unusual promoter. J Immunol 136: 1482–1489
Katoh K, Takahashi Y, Hayashi S, Kondoh H (1987) Improved mammalian vectors for high expression of G418 resistance. Cell Struct Funct 12:575–580
Martin GR (1980) Teratocarcinomas and mammalian embryogenesis. Science 209: 768–776
Morris R (1985) *Thy-1* in developing nervous tissue. Dev Neurosci 7:133–160
Muramatsu T (1988) Developmentally regulated expressin of cell surface carbohydrates during mouse embryogenesis. J Cell Biochem 36:1–14
Rettig WJ, Old LJ (1989) Immunogenetics of human cell surface differentiation. Annu Rev Immunol 7:481–511
Rickles RJ, Darrow AL, Strickland S (1989) Differentiation-responsive elements in the 5' region of the mouse tissue plasminogen activator gene confer two-stage regulation by retinoic acid and cyclic AMP in teratocarcinoma cells. Mol Cell Biol 9:1691–1704
Rousset J-P, Jami J, Dubois P, Aviles D, Ritz E (1980) Developmental potentialities and surface antigens of mouse teratocarcinoma x lymphoid cell hybrids. Somatic Cell Genet 6:419–433
Seki T, Spurr N, Obata F, Goyert S, Goodfellow P, Silver J (1985) The human *Thy-1* gene: structure and chromosomal location. Proc Natl Acad Sci USA 82:6657–6661
Solter D, Knowles BB (1978) Monoclonal antibody defining a stage-specific mouse embryonic antigen (SSEA-1). Proc Natl Acad Sci USA 75:5565–5569
Stern PL, Martin GR, Evans MJ (1975) Cell surface antigens of clonal teratocarcinoma cells at various stages of differentiation. Cell 6:455–465
Strickland S, Smith KK, Marotti KR (1980) Hormonal induction of differentiation in teratocarcinoma stem cells: generation of parietal endoderm by retinoic acid and dibutyryl cAMP. Cell 21:347–355
Tse AGD, Barclay AN, Watts A, Williams AF (1985) A glycophospholipid tail at the carboxyl-terminus of the *Thy-1* glycoprotein of neurons and thymocytes. Science 230:1003–1008
Williams AF, Gagnon J (1982) Neuronal cell Thy-1 glycoprotein: homology with immunoglobulin. Science 216:696–703
Young PR, Tilghman SM (1984) Induction of α-fetoprotein synthesis in differentiating F9 teratocarcinoma cells is accompanied by a genome-wide loss of DNA methylation. Mol Cell Biol 4:898–907

# Chromosomal Studies

## *Cytogenetic Studies of Testicular Germ Cell Tumors: Pathogenetic Relevance*

S.M.M.J. Castedo, J.W. Oosterhuis, and B. de Jong

Department of Medical Genetics, Medical Faculty of Porto, Hospital S. João, 4200 Porto, Portugal

### Introduction

Testicular germ cell tumors (TGCTs) are a heterogeneous group of neoplasms (see Mostofi et al. 1987; Ulbright and Roth 1987 for recent reviews). TGCTs of adults can be divided clinically and morphologically into two main entities: seminomas and nonseminomatous germ cell tumors (NSGCTs) (Mostofi and Sobin 1977; Mostofi 1980). Seminomas are composed of neoplastic germ cells sharing many histological, immunohistochemical, and ultrastructural characteristics with carcinoma in situ (CIS) cells of the testis (Jacobsen and Nørgaard-Pedersen 1984). NSGCTs are composed of neoplastic embryonic tissues (embryonal carcinoma, immature teratoma, and mature teratoma) and/or extraembryonic tissues (yolk sac tumor and choriocarcinoma). TGCTs with a seminoma and an NSGCT component are called combined tumors according to the British classification (Pugh 1976). Seminomas are less aggressive than NSGCTs as a group (although the aggressiveness of NSGCTs depends on the histological subtype, in particular the presence of embryonal carcinoma, yolk sac tumor, and/or choriocarcinoma).

Carcinoma in situ is the precursor lesion of most TGCTs of adults and is often found in the testicular parenchyma surrounding the invasive cancer.

### Pathogenetic Models

In essence there are two theories about the pathogenetic relationship between seminomas and NSGCTs (see Damjanov 1989; Oosterhuis et al. 1989 for review). One model favors independent origins of these two entities via CIS (Mostofi et al. 1987; Pierce and Abell 1970; Nochomowitz and Rosai 1978). The other model (Ewing 1911; Friedman 1951; Raghavan et al. 1982; Oliver 1987) suggests a single origin for these tumors with seminoma as a stage after CIS through which all TGCTs of adults [with the possible exception of spermatocytic seminoma (Müller et al. 1987)] progress.

Recent Results in Cancer Research, Vol. 123

The question concerning the possible pathogenetic relationship of seminomas and NSGCTs can be tackled in different ways. As will be shown, cytogenetics and DNA flow cytometry are powerful tools for the clarification of the above-mentioned problem.

## Ploidy Studies

Recent DNA flow cytometric studies of TGCTs of adults have shown consistent aneuploid peaks in the majority of tumors (Oosterhuis et al. 1989). The same studies have demonstrated that seminomas have a significantly higher DNA index (DI) (slightly hypertriploid) than NSGCTs and combined tumors (slightly hypotriploid). The DI of the seminoma component of combined tumors is also significantly higher than the DI of the nonseminomatous part.

The mean DI of CIS in patients who have not yet developed invasive cancer is peritetraploid, as was shown by Müller and Skakkebaek (1981).

Based on these findings, Oosterhuis et al. (1989) suggested that polyploidization of a dysplastic germ cell precursor resulting in a tetraploid CIS was probably an early event in the oncogenesis of TGCT. Conceivably, CIS might progress to invasive seminoma through a net loss of DNA. Further loss of chromosomes would result in an NSGCT, a more advanced cancer in terms of tumor evolution, which has lost the capability of gonocytic differentiation.

Cytogenetic studies of TGCTs are of crucial importance in elucidating whether the consistently lower DI of NSGCTs as compared with seminomas is due to nonrandom loss of certain critical (parts of) chromosomes.

## Cytogenetic Studies

Early cytogenetic studies of TGCTs described a generally hyperdiploid to hypertriploid chromosome complement with higher modal chromosome numbers in seminomas as compared with NSGCTs, and combined tumors with intermediate modes (Martineau 1966, 1969; Galton et al. 1966; Fischer and Golob 1967; Atkin 1973).

Recent studies confirmed the previous observations about modal chromosome numbers in TGCTs (see Castedo et al. 1989a,c for review).

The majority of reported seminomas show chromosome numbers around triploid and tetraploid values (see Castedo et al. 1989a for review), whereas most NSGCTs have between 60 and 64 chromosomes (see Castedo et al. 1989c for review). A cytogenetic study of three cases of CIS of the testis has recently been described (Vos et al. 1990), showing strongly aneuploid karyotypes in the peritriploid range and fewer structural abnormalities than the corresponding invasive tumors.

Both seminomas and NSGCTs show an apparent nonrandom gain and/or loss of certain chromosomes. In seminomas normal copies of chromosomes

11, 13, and 18 are underrepresented compared with the normal copies of chromosomes 7, 8, 15, 21, and X (Castedo et al. 1989a). In two different samples of NSGCTs, normal copies of chromosomes 11, 13, and 18 were consistently underrepresented compared with the normal copies of chromosomes 7, 8, 12, and X (Castedo et al. 1989c; Vos et al., unpublished data). It is conceivable that chromosomes consistently underrepresented in both subtypes (e.g., 11, 13, and 18) harbor genes important for normal germ cell differentiation and/or with tumor-suppressing properties. Chromosomes overrepresented (e.g., 7, 8, 12, and X) may contain genes responsible for a more malignant development.

The most common chromosomal structural abnormality found in TGCTs is the isochromosome 12p (Atkin and Baker 1983; Delozier-Blanchet et al. 1985). This marker is present in one or more copies in over 80% of all TGCTs (Castedo et al. 1988b) and is characteristic for all histological varieties of germ cell tumors of the testis (Delozier-Blanchet et al. 1987; Gibas et al. 1986). Its occurrence in the various histological types of TGCTs points to the oncogenetic importance of the i(12p) and to the pathogenetic interrelationship between the different subtypes of TGCTs. However, molecular studies carried out on fresh tumors and on TGCT-derived cell lines indicate that the formation of an i(12p) is probably not the first genetic event in the oncogenesis of TGCTs (Geurts van Kessel et al. 1989). The fact that in two out of three karyotyped CIS no i(12p) could be found lends further support to this hypothesis (Vos et al. 1990). The average number of copies of i(12p) in NSGCTs is significantly higher than in seminomas, which suggests a correlation with increased aggressiveness (Delozier-Blanchet et al. 1987; Castedo et al. 1989b). Several other, not consistent, structural abnormalities have been described in TGCTs.

Several hypotheses have been proposed on the mechanism of origination of aneuploidy in TGCTs: successive nondisjunctions of a diploid cell, polyploidization, and cell fusion, followed and/or preceded by gain and/or loss of specific chromosomes (see Sandberg 1980 for review). Since the DI of CIS is about 2 (Müller and Skakkebaek 1981), and the modal chromosome numbers of seminomas and NSGCTs clearly cluster around the triploid range, it seems rather unlikely that successive random nondisjunctions in a diploid cell could be a frequent oncogenetic mechanism of TGCTs. Thus, mitotic or meiotic errors leading to tri- or tetraploidization or cell fusion, followed by net loss of chromosomes, are more plausible hypotheses to explain the origin of the polyploid cell from which TGCTs derive.

If chromosomal loss in TGCTs is related to loss of genes crucial for normal germ cell differentiation, different chromosomes should be underrepresented in NSGCTs compared with seminomas. However, since both are germ cell tumors, one would also expect some similarities between the two subtypes.

The comparison of the average number of copies of the different chromosomes in seminomas and NSGCTs shows a significant similarity in their relative proportions, mainly for the nonacrocentric chromosomes (de Jong et al. 1990). Moreover, in the two published cases of a combined tumor in which both components were separately karytoyped using banding techniques (Haddad et al. 1988; Castedo et al. 1988a), there were clonal abnormalities common to both components, although in one case (Castedo et al. 1988a) the only structural abnormality common to both components was the i(12p). These findings are strong arguments in favor of a common origin and relationship of seminomas and NSGCTs. However, compared with seminomas, some acrocentric chromosomes (esp. #15 and #22) are significantly less represented in NSGCTs (Castedo et al. 1989a,c). It might be speculated that loss of these chromosomes is crucial for a seminoma stage cell to become a NSGCT, possibly because of loss of genes important for germ cell differentiation.

## Conclusions

The existing data concerning chromosomal studies and ploidy of TGCTs fit well in the model of pathogenesis of TGCTs proposed by Ewing (1911) and Friedman (1951), according to which seminomas and NSGCTs have a single origin and neoplastic pathway, with seminomas representing an intermediate stage in the development of NSGCT components.

Since CIS is peritetraploid in patients who have not yet developed invasive cancer (Müller and Skakkebaek 1981; Müller et al. 1987) and invasive TGCTs are usually peritriploid (Oosterhuis et al. 1989), it seems logical to assume that tumor progression in TGCTs is the result of a combination of chromosomal events leading to a net loss of DNA. This hypothesis is in disagreement with the tumor progression model proposed by Nowell (1986), according to which the clonal evolution of a tumor cell population goes from diploid to higher chromosome numbers. This decrease is probably the end result of the loss of specific chromosomes.

Thus, tumor progression of TGCTs will probably consist of the following steps, although not necessarily in this order:

1. Aneuploidy due to polyploidization or cell fusion
2. Genetic instability due to aneuploidy (Nowell 1986; Wolman 1986)
3. Development of structural chromosomal abnormalities, especially formation of i(12p), conceivably leading to (onco) gene deregulation
4. Gain or retention of chromosomes or chromosomal material (namely from chromosomes 7, 8, 12, and X) leading to a selective growth advantage
5. Loss of chromosomes or chromosomal material (namely from chromosomes 11, 13, and 18) leading to a selective growth advantage, either by loss of tumor suppression or loss of the ability for terminal differentiation (see Harris 1986; Sager 1986 for review).

*Acknowledgements.* This work was supported by the Netherlands Cancer Foundation grants GUCK 84–6 and 88–10, and partly by the J.K. de Cock Stichting, Groningen.

***References***

Atkin NB (1973) High chromosome numbers of seminomata and malignant teratoma of the testis: a review of data on 103 tumours. Br J Cancer 28:275–279
Atkin NB, Baker MC (1983) i(12p): Specific chromosomal marker in seminoma and malignant teratoma of the testis? Cancer Genet Cytogenet 10:199–204
Castedo SMMJ, de Jong B, Oosterhuis JW, Seruca R, Buist J, Schraffordt Koops H (1988a) Cytogenetic study of a combined germ cell tumor of the testis. Cancer Genet Cytogenet 35:159–165
Castedo SMMJ, de Jong B, Oosterhuis JW, Seruca R, Idenburg VJS, Buist J, Sleijfer DT (1988b) i(12p) negative testicular germ cell tumors. A different group? Cancer Genet Cytogenet 35:171–178
Castedo SMMJ, de Jong B, Oosterhuis JW, Seruca R, te Meerman GJ, Dam A, Schraffordt Koops H (1989a) Cytogenetic analysis of ten human seminomas. Cancer Res 49:439–443
Castedo SMMJ, de Jong B, Oosterhuis JW, Idenburg VJS, Seruca R, Buist J, te Meerman G, Schraffordt Koops H, Sleijfer DT (1989b) Chromosomal changes in mature residual teratomas following polychemotherapy. Cancer Res 49:672–676
Castedo S, de Jong B, Oosterhuis JW, Seruca R, Idenburg VJS, Dam A, te Meerman GJ, Schraffordt Koops H, Sleijfer DT (1989c) Chromosomal changes in primary testicular nonseminomas. Cancer Res 49:5696–5701
Damjanov I (1989) Is seminoma a relative or a precursor of embryonal carcinoma? Lab Invest 60:1–3
De Jong B, Oosterhuis JW, Castedo SMMJ, Vos AM, te Meerman GJ (1990) Pathogenesis of adult testicular germ cell tumors. A cytogenetic model. Cancer Genet Cytogenet 48:143–167
Delozier-Blanchet CD, Engel E, Walt H (1985) Isochromosome 12p in malignant testicular tumors. Cancer Genet Cytogenet 15:375–376
Delozier-Blanchet CD, Walt H, Engel E, Vaugnat P (1987) Cytogenetic studies of human testicular germ cell tumors. Int J Androl 10:69–78
Ewing J (1911) Teratoma testis and its derivatives. Surg Gynecol Obstet 12:230–261
Fischer P, Golob E (1967) Similar marker chromosomes in testicular tumours. Lancet 1:216
Friedman NB (1951) The comparative morphogenesis of extragenital and gonadal teratoid tumors. Cancer 4:265–276
Galton M, Benirschke K, Baker M, Atkin NB (1966) Chromosomes of testicular teratomas. Cytogenetics 6:261–275
Geurts van Kessel A, van Drunen E, de Jong B, Oosterhuis JW, Langeveld A, Mulder MP (1989) Chromosome 12q heterozygosity is retained in i(12p) positive testicular germ cell tumors. Cancer Genet Cytogenet 40:129–134
Gibas Z, Prout GR, Pontes JE, Sandberg AA (1986) Chromosome changes in germ cell tumors of the testis. Cancer Genet Cytogenet 19:245–252
Haddad FS, Sorini PM, Somsin AA, Nathan MH, Dobbs RM, Berger CS, Sandberg AA (1988) Familial double testicular tumors: identical chromosome changes in seminoma and embryonal carcinoma of the same testis. J Urol 139:748–750
Harris H (1986) The genetic analysis of malignancy. J Cell Sci [Suppl] 4:431–444
Jacobsen GK, Nørgaard-Pedersen B (1984) Placental alkaline phosphatase in testicular germ cell tumors and in carcinoma-in-situ of the testis. Acta Pathol Microbiol Immunol Scand [A] 92:323–329
Martineau M (1966) A similar marker chromosome in testicular tumors. Lancet 1:839–842
Martineau M (1969) Chromosomes in human testicular tumours. J Pathol 99: 271–282
Mostofi FK (1980) Pathology of germ cell tumors of testis. Cancer 45:1735–1754

Mostofi FK, Sobin LH (1977) International histological classification of testicular tumors. WHO Geneva (International histologic classification of tumors, no 16)
Mostofi FK, Sesterhenn IA, Davis CJ Jr (1987) Immunopathology of germ cell tumors of the testis. Semin Diagn Pathol 4:320–341
Müller J, Skakkebaek NE (1981) Microspectrophotometric DNA measurements of carcinoma-in-situ germ cells in the testis. Int J Androl [Suppl] 4:211–221
Müller J, Skakkebaek NE, Parkinson MC (1987) The spermatocytic seminoma: views on pathogenesis. Int J Androl 10:147–156
Nochomowitz LE, Rosai J (1978) Current concepts on the histogenesis, pathology and immunohistochemistry of germ cell tumors of the testis. Pathol Annu 13:327–362
Nowell PC (1986) Mechanisms of tumor progression. Cancer Res 46:2203–2207
Oliver RTD (1987) HLA phenotype and clinicopathological behaviour of germ cell tumours: possible evidence for clonal evolution from seminomas to nonseminomas. Int J Androl 10:85–94
Oosterhuis JW, Castedo SMMJ, de Jong B, Cornelisse CJ, Dam A, Sleijfer DT, Schraffordt Koops H (1989) Ploidy of primary germ cell tumors of the testis. Pathogenetic and clinical relevance. Lab Invest 60:14–21
Pierce GB, Abell MR (1970) Embryonal carcinoma of the testis. Pathol Annu 5:27–60
Pugh RCB (1976) Combined tumours. In: Pugh RCB (ed) Pathology of the testis. Blackwell, Oxford, pp 245–258
Raghavan D, Sullivan AL, Peckham MJ, Neville AM (1982) Elevated serum alphafetoprotein and seminoma. Clinical evidence for a histologic continuum? Cancer 50:982–989
Sager R (1986) Genetic suppression of tumor formation: a new frontier in cancer research. Cancer Res 46:1573–1580
Sandberg AA (1980) The chromosomes in human cancer and leukemia. Elsevier, Amsterdam pp 511–515
Ulbright TM, Roth LM (1987) Recent developments in the pathology of germ cell tumors. Semin Diagn Pathol 4:304–319
Vos AM, Oosterhuis JW, de Jong B, Buist J, Schraffordt Koops H (1990) Cytogenetics of carcinoma in situ of the testis. Cancer Genet Cytogenet 46:75–81
Wolman SR (1986) Cytogenetic heterogeneity: its role in tumor evolution. Cancer Genet Cytogenet 19:129–140

# *Isochromosome (12p) in Germ Cell Tumors*

P. Dal Cin and H. Van Den Berghe

Center for Human Genetics, Herestraat 49, 3000 Leuven, Belgium

## Introduction

Germ cell tumors (GCTs) are benign or malignant neoplasms, presumably derived from pluripotent or primordial germ cells. During the development of the embryo, germ cells migrate from the yolk sac endoderm usually to the genital ridge in the retroperitoneum, where the gonads develop (Patton 1953). Aberrant migration, however, can also occur, and GCTs can arise in extragonadal sites, mostly in the anterior mediastinum (Gonzalez-Crussi 1982). Histologically the same subtypes exist regardless of whether the tumors are located in the gonads or elsewhere. There is a difference in the incidence of malignancy, which is relatively low among ovarian GCTs and very high among testicular GCTs (TGCTs) (Talerman 1985).

## Chromosome Changes in GCTs with Special Reference to i(12p)

### *Testicular Germ Cell Tumors*

As a rule, TGCT karyotypes are near-triploid or near-tetraploid, with several numerical and structural chromosome changes. Among these the occurrence of one or more copies of an isochromosome of the short arm of chromosome 12, i(12p), is striking and apparently characteristic of 80% of this kind of tumor of the testis (Castedo et al. 1988) (Fig. 1).

### *Ovarian Germ Cell Tumors*

Little information is available on chromosome changes in ovarian GCT (Speleman et al. 1990). No specific chromosome change seems to be associated, but four recent reports describe the presence of one or more copies of

Recent Results in Cancer Research, Vol. 123

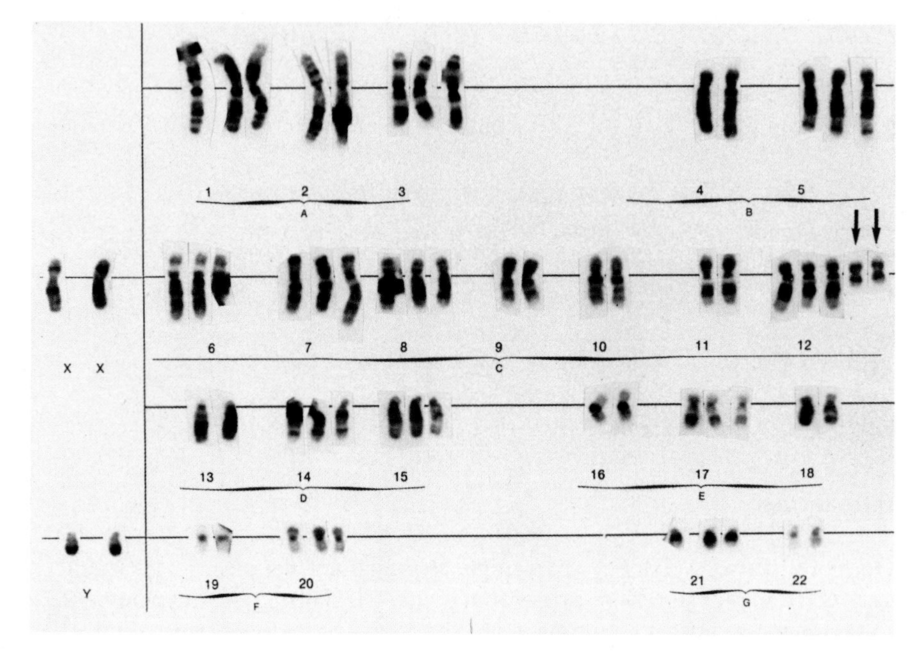


**Fig. 1.** Representative G-banded karyotype from a TGCT showing two extra copies of i(12p)

i(12p) in two dysgerminomas, one malignant mixed Müllerian tumor and one yolk sac tumor (Atkin and Baker 1987; Jenkyn and McCartney 1987; Speleman et al. 1990) (Fig. 2).

### *Extragonadal Germ Cell Tumors*

So far only five cytogenetic analyses on extragonadal GCTs, all mediastinal tumors, have been reported (Kaplan et al. 1979; Oosterhuis et al. 1985; Mann et al. 1983; Dal Cin et al. 1989; Chaganti et al. 1989). Of these, two, perhaps three cases showed the presence of only one extra copy of i(12p) in a near-diploid karyotype. We reported a 47, XY, −13, +21, +i(12p) karyotype in the primary tumor, with four distinct histological components such as differentiated teratoma, embryonal carcinoma, seminoma, and yolk sac tumor (Dal Cin et al. 1989) (Fig. 3). Chaganti et al. (1989) reported a 48–49, XY, +1, +6, +i(12p) karyotype in a primary tumor, with a yolk sac tumor component as well as an immature teratoma.

A mediastinal embryonal cell carcinoma, in a patient with Klinefelter's syndrome, cytogenetically investigated by Mann et al. (1983), showed a karyotype with 50, XXY, +7, +21, +mar. The marker chromosome was interpreted as an isochromosome of the long arm of chromosome 21, but it

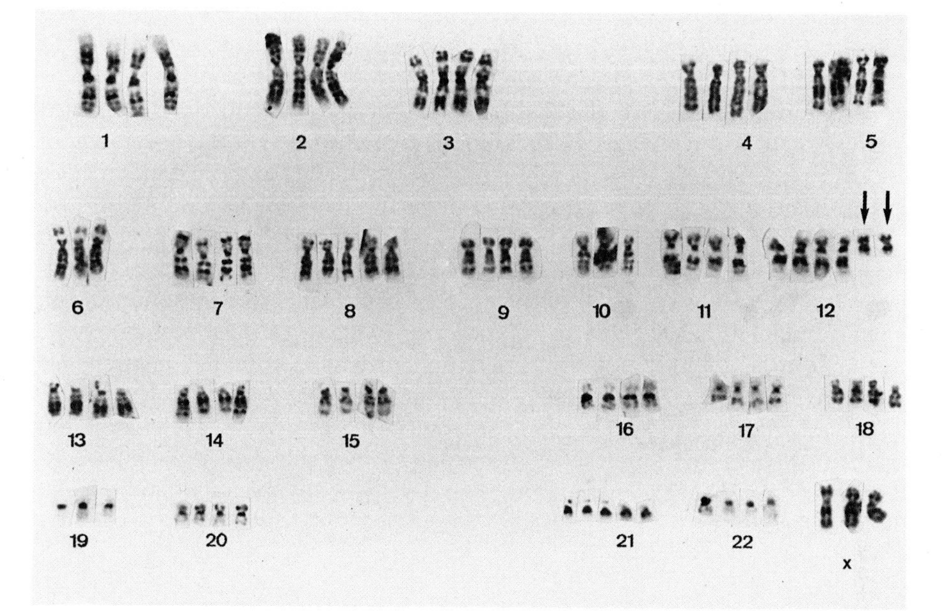


**Fig. 2.** Representative G-banded karyotype from an ovarian germ cell tumor, with 92 chromosomes and 2 extra copies of i(12p)

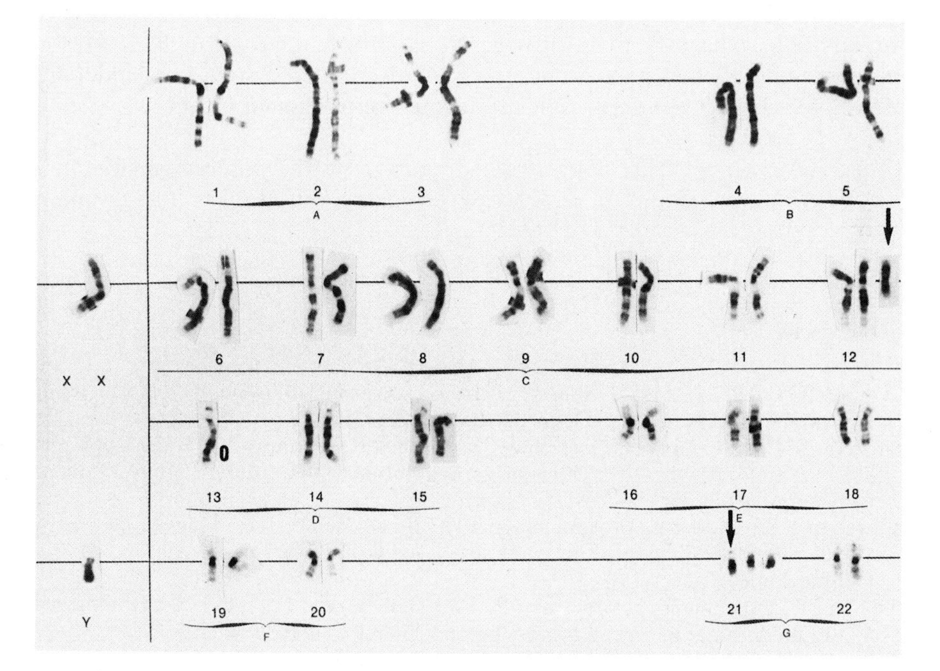


**Fig. 3.** Representative G-banded karyotype from a mediastinal germ cell tumor showing one extra copy of i(12p)

may have been an i(12p). Whether the gain of an extra chromosome 21 may also be a nonrandomly occurring chromosomal change in mediastinal GCTs must await further investigation.

Two observations are particularly relevant to extragonadal GCTs: (a) patients with Klinefelter's syndrome may be predisposed to the development of GCTs, particularly those of extragonadal origin and (b) there is an intrinsic relationship between hematological malignancies and mediastinal GCTs. So far only one case of mediastinal GCT in a patient with Klinefelter's syndrome has been cytogenetically investigated (Mann et al. 1983). A possible common biological basis for the occurrence of acute leukemia in patients with mediastinal GCT has been a matter of speculation (Nichols et al. 1985; Dement et al. 1985) because no single patient having both malignancies has been investigated cytogenetically, since the recent report of Chaganti et al. (1989). An extra copy of i(12p) and an extra chromosome 21 have been found in the postchemotherapy resected tumor and in two subsequent bone marrow aspirates (ANLL-M2). These findings tend to show an intrinsic relationship of both malignancies.

## Conclusion

Despite the paucity of cytogenetic investigations in GCT females and in extragonadal GCTs, in contrast to extensive reports on TGCTs, the presence of an i(12p) suggests that through an unknown mechanism this specific chromosome change may occur in an undifferentiated stem cell, independently of whether the germ cells are located in the gonad or not.

*Acknowledgement.* This study was supported by the Interuniversity Network for Fundamental Research sponsored by the Belgian Government (1987–1991).

## *References*

Atkin NB, Baker MC (1987) Abnormal chromosomes including small metacentrics in 14 ovarian cancers. Cancer Genet Cytogenet 26:355–361

Castedo SMMJ, de Jong B, Oosterhuis JW, Seruca R, Idenburg VJ, Buist J, Sleijfer DT (1988) i(12p)-Negative testicular germ cell tumors. A different group? Cancer Genet Cytogenet 35:171–178

Chaganti RSK, Ladanyi M, Samaniego F, Offit K, Reuter VE, Jhanwar SC, Bosl GJ (1989) Leukemic differentiation of a mediastinal germ cell tumor. Genes Chromosomes Cancer 1:83–87

Dal Cin P, Drochmans A, Moerman P, Van Den Berghe H (1989) Isochromosome 12p in mediastinal germ cell tumor. Cancer Genet Cytogenet 42:243–251

Dement SH, Eggleston JC, Spivak JL (1985) Association between mediastinal germ cell tumors and hematologic malignancies. Report of two cases and review of the literature. Am J Surg Pathol 9:23–30

Fox RM, Woods RL, Tattersall MHN (1979) Undifferentiated carcinoma in young men: the atypical teratoma syndrome. Lancet 1:1316–1318

Gonzales-Crussi F (1982) Extragonadal teratomas. In: AFIP (ed) Atlas of tumor pathology. Armed Forces Institute of Pathology, Washington

Jenkyn DJ, McCartney AJ (1987) A chromosome study of three ovarian tumors. Cancer Genet Cytogenet 26:327–337

Kaplan CG, Askin FB, Benirschke K (1979) Cytogenetics of extragonadal tumors. Teratology 19:261–266

Mann BD, Sparkes RS, Kern DH, Morton DL (1983) Chromosomal abnormalities of a mediastinal embryonal cell carcinoma in a patient with 47, XXY Klinefelter syndrome: evidence for the premeiotic origin of a germ cell tumor. Cancer Genet Cytogenet 8:191–196

Nichols CR, Hoffman R, Einhorn LH, Williams SD, Wheeler LA, Garnick MB (1985) Hematologic malignancies associated with primary mediastinal germ-cell tumors. Ann Intern Med 102:603–609

Oosterhuis JW, de Jong B, van Dalen I, van der Meer I, Visser M, de Leij L., Mesander G, Collard JG, Schraffordt Koops H, Sleijfer DT (1985) Identical chromosome translocations involving the region of the c-*myb* oncogene in four metastases of a mediastinal teratocarcinoma. Cancer Genet Cytogenet 15:99–107

Patton BM (1953) Human embryology, (2nd edn.) Blakiston, New York

Speleman F, De Potter C, Dal Cin P, Mangelschots K, Ingelaere H, Laureys G, Benoit Y, Leroy J, Van Den Berghe H (1990) i(12p) in malignant ovarian tumor. Cancer Genet Cytogenet 45:49–53

Talerman A (1985) Germ cell tumors and mixed germ cell – sex cord – stromal tumors of the ovary. In: Roth LM, Czernobilsky B, (eds) Tumors and tumorlike conditions of the ovary. Churchill Livingstone, New York, p 75

Molecular Genetics

# *Molecular Analysis of Isochromosome 12p in Testicular Germ Cell Tumors*

A. Geurts van Kessel,[1] R. Suijkerbuijk,[1] B. de Jong,[2] and J.W. Oosterhuis[2]

[1] Department of Human Genetics, University Hospital, PO Box 9101, 6500 HB Nijmegen, The Netherlands
[2] Departments of Human Genetics and Pathology, University of Groningen, Groningen, The Netherlands

## Introduction

Testicular germ cell tumors (TGCTs) of adults are marked by a common structural chromosomal abnormality: i(12p) (Atkin and Baker 1988). This chromosome anomaly is characteristic for all histologic varieties of TGCTs and is observed in more than 80% of all cases. This frequent occurrence suggests that the altered chromosome 12 could play an important role in the pathogenesis of germ cell tumors. However, at the present time little is known about the role of this chromosome in tumorigenesis. In addition, TGCTs are usually highly aneuploid. Thus, besides i(12p) formation, aneuploidization could play an important role in the oncogenesis of TGCTs. We supply evidence suggesting that i(12p) formation is not the first step in the development of TGCTs and that its generation does not lead to gross loss of material of the long arm of chromosome 12. We also show that in situ hybridization ("chromosome painting") represents a valuable new approach for the detection of the i(12p) chromosome in germ cell tumor material.

## Materials and Methods

High molecular weight DNAs were extracted from primary tumor samples, normal tissues, and cultured cell lines as reported previously (Geurts van Kessel et al. 1985). DNAs were cleaved to completion with restriction enzymes according to the specification of the suppliers, subjected to electrophoresis through 0.7% agarose gels, and transferred to Nytran filters (New England Nuclear). Probe labeling was carried out using random priming (Feinberg and Vogelstein 1983). After hybridization, filters were washed to 0.3 × sodlium saline citrate (SSC) at 65°C. Fuji X-ray films were exposed at −80°C for 1–3 days using intensifying screens. Nonradioactive in situ hybridizations were carried out essentially as described by Pinkel et al. (1986).

Recent Results in Cancer Research, Vol. 123

Total human or hybrid cell DNA was labeled with biotin-11-dioxyuridine triphosphate (dUTP) or biotin-16-dUTP by nick translation. Hybridizations were carried out in a moist chamber at 37°C for 24 h. Detection of the biotinylated probes was achieved using fluorescein isothiocyanate (FITC) – conjugated avidin followed by an amplification step using goat antiavidin antibodies and again FITC-conjugated avidin (Vector laboratories). After washing, the slides were mounted in antifade medium (DABCO, Sigma) containing 4,6-diamino-2-phenylindole (DAPI, Sigma) for counter-staining of the chromosomes.

## Results and Discussion

In general, the formation of an isochromosome in diploid cells leads to loss of the chromosomal arm not included in the anomaly. As a consequence, this leads to a loss of heterozygosity of genes located on the deleted chromosomal arm. In the case of the i(12p) chromosome, this results in the loss of heterozygosity of genes on the q arm of chromosome 12. Subsequent aneuploidization may generate multiple identical copies of the unaffected chromosome 12 in the tumor cells. Alternatively, if aneuploidization occurs prior to generation of the i(12p) marker, this sequence of events will not necessarily result in the loss of heterozygosity of genes located on 12q.

To test the order of these events, we assayed TGCT cells for the presence or absence of chromosome 12q heterozygosity using restriction fragment length polymorphism (RFLP) analysis. For this analysis we used various primary tumor samples and i(12p)-positive cell lines (Andrews et al. 1986; Geurts van Kessel et al. 1989). In Fig. 1, examples of the RFLP analyses are presented. Screening of normal DNAs obtained from several TGCT patients for the presence of heterozygous alleles of the polymorphic insulin-like growth factor-1 (*IGF-1*) locus on chromosome 12q after *Pvu*II digestion (Höppener et al. 1985) revealed that patients A, D, E, and F were negative and that patients B and C were positive and, therefore, informative for our present study (5.1- and 4.7-kb bands). Subsequent analysis of the tumor DNAs from seminoma and embryonal carcinoma clearly revealed the retention of this heterozygosity (Fig. 1, lanes B, C). Previously, we had already demonstrated that TGCT-derived NT2/D1 cells are heterozygous for the polymorphic 12q markers D12S4 using *Taq*1-cleaved DNA and D12S6 and D12S8 using *Msp*1-cleaved DNAs (Geurts van Kessel et al. 1989). In the same study, very similar results were obtained with another TGCT line, Scha-1. Moreover, via somatic cell hybrid analysis, we were able to demonstrate in the tumor cells the presence of at least two intact (maternal and paternal derived) allelic copies of chromosome 12 (Geurts van Kessel et al. 1989).

Taken together, we conclude that loss of gene heterozygosity on the q arm of chromosome 12 is not a general characteristic of i(12p)-positive TGCTs.

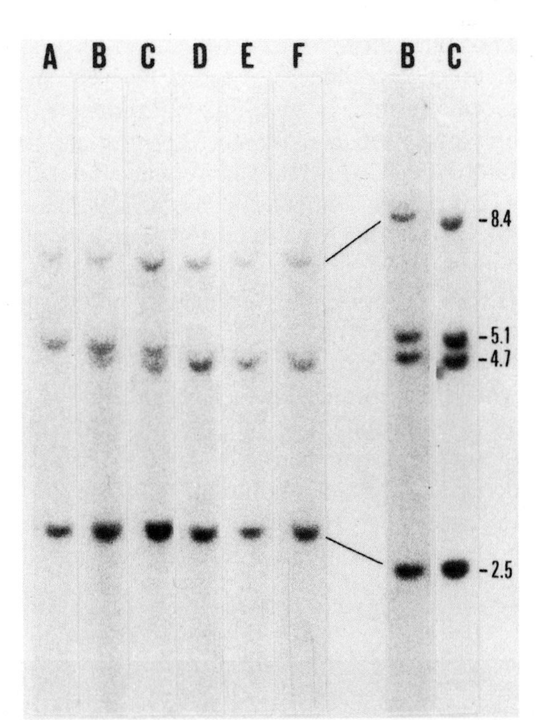


**Fig. 1.** Restriction fragment length polymorphism analysis of normal (*left panel*) and TGCT-derived (*right panel*) DNAs cleaved with *Pvu*II and hybridized with an *IGF-1* probe. The 5.1- and 4.7-kb hybridizing fragments represent different polymorphic alleles. The 8.4- and 2.5-kb bands represent constitutively hybridizing fragments (7). Patients *B* and *C* are heterozygous in both normal (*left*) and tumor-derived (*right*) samples

Therefore, teleologically, we argue that aneuploidization seems to be at least one step ahead of the formation of the i(12p) anomaly during the evolution of these tumors. This conclusion is in agreement with the previous suggestion of Oosterhuis et al. (1988) that polyploidization is an early step in the pathogenesis of germ cell tumors of the testis. Our results are also in accordance with the suggestion made by Delozier-Blanchet et al. (1987) that multiple copies of the i(12p) marker usually characterize tumors with aggressive growth and therefore in advanced clinical stages at the tissue of diagnosis. The exact nature of the i(12p) anomaly has, so far, not been determined. Using somatic cell hybrid analysis we have previously demonstrated that the i(12p) marker is indeed derived from the short arm of chromosome 12 (Geurts van Kessel et al. 1989). Recently, a new approach has been developed that will enable us (a) to reveal the exact nature of the TGCT-specific marker and (b) to determine in detail its occurrence during the etiology of germ cell tumors. This new approach is based on already existing non-radioactive in situ hybridization procedures which allow visualization of a whole chromosome or a specific fragment thereof. For these "chromosome painting" procedures, various probes can be used such as chromosome-specific cosmid, or phage libraries, or somatic cell hybrids that have retained only one particular human chromosome (Pinkel et al. 1988; Cremer et al. 1988). A cell hybrid that has retained the Pallister-Killian syndrome-associated i(12p) chromosome has been isolated (Zhang et al. 1989). Molecular analysis has shown that this hybrid indeed contains only 12p material. When we hybridized metaphase chromosomes from this cell line with total human DNA as a probe, only a single human chromosome with i(12p) morphology could be observed (Fig. 2), often present in multiple copies per cell. When, in turn, DNA extracted from this hybrid was used as a probe on normal human metaphase chromosomes, hybridization with 12p was observed exclusively, again confirming that only 12p material is present. Obviously, application of this new probe, in conjunction with "chromosome painting", will enable us to determine the exact nature of the i(12p) chromosome in TGCTs. Moreover, since these procedures can in principle also be applied to interphase nuclei, in both fresh and archival material, we may now have the tools to study the occurrence of this specific marker during the etiology of TGCT.

**Fig. 2.** Visualization of the Pallister-Killian-derived i(12p) chromosome in hybrid cell line M28 (Zhang et al. 1989). Hybridization was carried out using biotin-labeled total human DNA. The i(12p) marker is present as the only human chromosome in multiple copies (*arrows*)

## Conclusion

We have demonstrated, by using RFLP analysis, that formation of the i(12p) chromosome in TGCTs does not lead to loss of heterozygosity of various loci on the q arm of chromosome 12. This result suggests that aneuploidization precedes the formation of the i(12p) marker chromosome during the malignant transformation leading to the formation of germ cell tumors. In situ hybridization ("chromosome painting") is a valuable new approach for the detection of the i(12p) chromosome in germ cell tumors.

*Acknowledgements.* The authors thank Drs. T. Hopman, D. Smeets, and H. Ropers for advice and support and Mrs. M. Ariaans for secretarial assistance. We acknowledge Drs. Cassiman and J. Höppener for providing cells and probes. This work was supported by the Netherlands Cancer Society (Koningin Wilhelmina Fonds).

## *References*

Andrews PW, Gonczol E, Plotkin SA, Dignazio M, Oosterhuis JW (1986) Differentiation of tera-2 human embryonal carcinoma cells into neurons and HCMV permissive cells. Differentiation 31:119–126

Atkin NB, Baker MC (1988) Small metacentric marker chromosomes, particularly isochromosomes, in cancer. Hum Genet 79:96–102

Cremer T, Lichter P, Borden I, Ward DC, Manuelides L (1988) Detection of chromosome aberrations in metaphase and interphase tumor cells by in situ hybridization using chromosome-specific library probes. Hum Genet 80:235–246

Delozier-Blanchet CD, Walt H, Engel E, Vuagnat P (1987) Cytogenetic studies of human testicular germ cell tumors. Int J Androl 10:69–77

Feinberg AP, Vogelstein B (1983) A technique for radiolabeling DNA restriction endonuclease fragments to high specific activity. Anal Biochem 132:6–13

Geurts van Kessel A, Turc-Carel C, de Klein A, Grosveld G, Lenoir G, Boostma D (1985) Translocation of oncogene c-*sis* from chromosome 22 to chromosome 11 in a Ewing sarcoma derived cell line. Mol Cell Biol 5:427–429

Geurts van Kessel A, van Drunen E, de Jong B, Oosterhuis JW, Langeveld A, Mulder MP (1989) Chromosome 12q heterozygosity is retained in i(12p)-positive testicular germ cell tumor cells. Cancer Genet Cytogenet 40:129–134

Höppener I, de Pagter-Holthuizen P, Geurts van Kessel A, Jansen M, Kittur S, Antonarakis S, Lips C, Sussenbach I (1985) The human gene encoding insulin-like growth factor 1 is located on chromosome 12. Hum Genet 69:157–160

Oosterhuis JW, Castedo SMMJ, de Jong B, Cornelisse CJ, Dam A, Sleijfer DT, Schraffordt Koops H (1988) Ploidy of subtypes of primary germ cell tumors of the testis. Pathogenic and clinical relevance. Lab Invest 60:14–21

Pinkel D, Straume T, Gray JW (1986) Cytogenetic analysis using quantitative, high sensitivity, fluorescence hybridization Proc Natl Acad Sci USA 83:2934–2938

Pinkel D, Landegent I, Collins C, Fuscoe J, Segraves R, Lucas J, Gray I (1988) Fluorescence in situ hybridization with human chromosome-specific libraries: detection of trisomy 21 and translocations of chromosome 4. Proc Natl Acad Sci USA 85:9238–9242

Zhang I, Marijnen P, Devriendt K, Frijns JP, van Den Berghe H, Cassiman JJ (1989) Molecular analysis of the isochromosome 12p in Pallister-Killian syndrome. Construction of a mouse-human cell line containing an i(12p) as the sole human chromosome. Hum Genet 83:359–363

# *Analysis of Chromosome 12 Abnormalities in Male Germ Cell Cancers*

E. Dmitrovsky, E. Rodriguez, F. Samaniego, V.E. Reuter, W.H. Miller, Jr., N.L. Geller, G.J. Bosl, and R.S.K. Chaganti

Laboratory of Molecular Medicine, Genitourinary Oncology Service, Memorial Sloan-Kettering Cancer Center, Cornell University Medical College, New York, NY 10021, USA

## Introduction

A specific cytogenetic abnormality, an isochromosome 12p, i(12p), has been described in male germ cell cancers (Atkin and Baker 1982, 1983). The incidence of this abnormality has been stated to be as high as 90% (Heim and Mitelman 1987), occurring in tumors histopathologically diagnosed as seminoma, nonseminoma, and teratoma (Castedo et al. 1989a,b,c; Bosl et al. 1989; Samaniego et al. 1990). Our investigations of i(12p) in human germ cell cancer provide further support for the diagnostic importance of this cytogenetic marker. This review presents the molecular and cytogenetic studies from the Memorial Sloan-Kettering Cancer Center principally concerning chromosome 12 in primary and metastatic germ cell tumors and in established germ cell cancer cell lines.

## Cytogenetic Abnormalities

A comprehensive cytogenetic analysis was performed by Chaganti and colleagues on primary gonadal, extragonadal, and metastatic germ cell tumors as well as on a number of established germ cell cancer cell lines (Chaganti et al. 1989; Ladanyi et al. 1990; Samaniego et al. 1990; Murty et al. 1990). An i(12p) was identified in all histologic subtypes (including choriocarcinoma and mature teratoma) and in tumors obtained from both gonadal and extragonadal primary sites (Bosl et al. 1989; Samaniego et al. 1990). In 29 tumor specimens obtained from 24 male germ cell tumors, an i(12p) was found in 20 specimens from 16 patients, representing an identification in 69% of all our examined specimens. Two specimens had a del (12q) in addition to the i(12p) and eight had a normal karyotype (46, XY), suggesting outgrowth of the nonmalignant elements within the tumors. One patient's tumor was a cytogenetic failure. Other cytogenetic alterations included the near-triploid

Recent Results in Cancer Research, Vol. 123

karyotypes in 15/16 (94%) patients, homogeneous-staining regions (HSRs) in 3/16 (19%) patients, and double minute chromosomes in 2/16 (13%) patients. Chromosomes 1, 7, 9, 12, 17, 21, 22, and X were nonrandomly gained in these tumors and a del(12) (q13–q22) was observed in 44% of nonseminomatous and mixed germ cell tumors (Samaniego et al. 1990). Taken together, these data along with those from the published literature provide evidence for the specificity of this cytogenetic change in germ cell tumors and raise the prospect of the importance of i(12p) in the transformation of normal human germ cells. Diagnostically, the examination of tumors of unclear histogenesis such as in the "unrecognized germ cell tumor syndrome" (Greco et al. 1986) for the presence of i(12p) might aid in their clinical management since tumors with an i(12p) may respond to germ cell tumor treatments.

One patient among these 24 presented with a mediastinal primary germ cell cancer. He was treated with a combination chemotherapy regimen, but relapsed with both progressive mediastinal disease and an acute myelomonocytic leukemia 11 months after the germ cell tumor diagnosis (Chaganti et al. 1989). Isochromosome (12p) was identified in the original mediastinal tumor specimen and in the acute leukemia, suggesting that the clonal origin of the acute leukemia cell and the germ cell tumor were the same. The germ cell origin of a subset of leukemias arising in patients with mediastinal germ cell tumors was confirmed on the basis of the presence of the i(12p) and/or common immunohistochemical markers (Ladanyi et al. 1990). These studies highlight the potential for differentiation which germ cell tumors can exhibit and the value of the i(12p) as a marker in germ cell derived tumors.

Since nearly all patients were entered onto ongoing germ cell tumor treatment protocols, studies of the correlation between 12p copy number and treatment outcome were undertaken. The presence of three or more additional copies of 12p was associated with a statistically greater likelihood of a clinical treatment failure (Bosl et al. 1989). This is analogous to the experience in chronic myelogenous leukemia in which the appearance of two copies of the Philadelphia chromosome is a harbinger of a more aggressive clinical phase (Watmore et al. 1985). In contrast, patients whose tumors had a normal karyotype or had two or less additional copies of 12p had a more favorable treatment outcome (Bosl et al. 1989). These data suggest that cytogenetics may provide important prognostic information, in addition to clinical features (Bosl et al. 1983) such as sites of metastases, lactate dehydrogenase (LDH) levels, and human chorionic gonadotropin (HCG) values which are known to be important in this malignancy.

In order to investigate chromosome 12 abnormalities in greater detail, cytogenetic studies of established human germ cell cancer cell lines were undertaken. An i(12p) marker was present in seven out of seven male germ cell cancer cell lines, but not in a single female teratocarcinoma cell line (Dmitrovsky et al. 1990). Other nonrandom chromosome 12 abnormalities contained within these male germ cell cancer cell lines included a del(12q14) in five out of seven lines (Murty et al. 1990). Other consistent nonrandom abnormalities included a del(1p22), del(1q21), i(1q), and del(7q11.2) (Murty et al. 1990). Thus, the investigation of germ cell tumor cell lines has demonstrated no discrepancy regarding the occurrence of i(12p) in fresh tumors versus established cell lines. These data suggest that, in addition to the 12p gain, the loss of 12q might also be important in the transformation process of germ cell tumors. Cell lines represent useful tools to examine genetic changes in a homogeneous germ cell tumor population.

## Molecular Genetic Studies

Given the common occurrence and the clinical importance of the i(12p) in primary and metastatic germ cell tumors, a more rapid approach to determine 12p copy number was needed. This would expedite the analysis of fresh tumors since cytogenetic studies are not always successful. As a first step in these studies, established germ cell cancer cell lines were used as a source of homogeneous germ cell tumors. Since these cell lines had been fully karyotyped (Murty et al. 1990), a molecular genetic determination of 12p copy number could be assessed as a possible substitute for routine cytogenetics. Using DNA probes for genes located on 12p (c-Ki-*ras*$_2$ and *D12S2*) and 12q (*int*-1), the densitometric intensities of Southern blot signals obtained for these genes within these tumors and in human placenta were compared. A reference gene, *bcl*-1, was selected since it maps to 11q, a chromosomal arm which exhibited infrequent numerical or structural changes in the examined germ cell tumors. This gene was also expected to have infrequent rearrangements in these nonhematopoietic tumors. The intensities of the obtained Southern blot signals for these genes were correlated with the known karyotypes. The Southern blotting data closely corroborated the cytogenetic studies (Dmitrovsky et al. 1990). This suggested that Southern blotting might substitute for routine cytogenetics when karyotypic studies cannot be performed.

To test this possibility, representative portions of male germ cell tumors were analyzed by Southern blotting using a modification of this described technique. Increased 12p copy number could also be identified in a subset of these tumors by Southern blotting (Samaniego et al. 1990). However, the presence of normal cells within these tumors diluted the contribution of the malignant cells. Perhaps with enrichment of the malignant elements of these tumors, the sensitivity of this Southern blot approach could be augmented.

More subtle structural changes involving chromosome 12 in germ cell cancers are possible. One example is the possible activation of the c-Ki-*ras*$_2$ gene, which is located on 12p. It has been suggested that enhanced expression or amplification of c-Ki-*ras*$_2$ might contribute to the transformation of germ cell tumors (Wang et al. 1987). Using a panel of established germ cell cancer cell lines, the amplification of c-Ki-*ras*$_2$ was found most likely to be the result of karyologic amplification (Dmitrovsky et al. 1990). Although enhanced expression of c-Ki-*ras*$_2$ mRNA was seen in conjunction with the i(12p), the polymerase chain reaction technique using specific mutated and nonmutated oligonucleotides for c-Ki-*ras*$_2$ codons 12, 13, or 61 (the commonly activated codons known to unleash the transforming potential of c-Ki-*ras*$_2$ protein) did not identify these mutations (Dmitrovsky et al. 1990). Since the level of c-Ki-*ras*$_2$ expression in the progenitor cells of germ cell tumors has not as yet been examined, we cannot exclude the possible contribution to germ cell transformation of the enhanced expression of this non-point-mutated gene. However, the point-mutational activation of c-Ki-*ras*$_2$ is at most an infrequent event in these nonseminomatous tumors. In contrast, in a subset of seminomatous tumors, activation of c-Ki-*ras*$_2$ has been reported (Mulder et al. 1989).

These data indicate that heterogeneity in c-Ki-*ras*$_2$ activation is possible in seminoma compared with nonseminomatous germ cell cancers. Subtle genomic changes involving chromosome 12 might also contribute to the transformation process or growth advantage of these germ cell tumor cells. This viewpoint is supported by the report of some germ cell tumors that lack the i(12p) (Castedo et al. 1988).

## Conclusion

Chromosome 12 abnormalities in germ cell cancers are frequent and are of likely clinical and biological importance. Further studies concerning genetic changes in germ cell cancer are clearly warranted. The challenge is to identify the gene or genes located on chromosome 12 or elsewhere in the genome which are centrally involved in the transformation process of normal human germ cells.

*Acknowledgements.* This study was supported in part by the American Cancer Society grant #PDT-381 and NCI grants and contracts CM-57332 and CA-05826, a Robert Wood Johnson Foundation Fellowship to Dr. Samaniego, a Lederle Fellowship to Dr. Miller, and an American Cancer Society Clinical Oncology Career Development Award #89-129 to Dr. Dmitrovsky.

## *References*

Atkin NB, Baker MC (1982) Specific chromosome change, i(12p) in testicular tumours? Lancet 2:1349

Atkin NB, Baker MC (1983) i(12p): Specific chromosome marker in seminoma and malignant teratoma of the testis? Cancer Genet Cytogenet 10:199–204

Bosl GJ, Geller NL, Cirrincione C, et al. (1983) Multivariate analysis of prognostic variables in patients with metastatic testicular cancer. Cancer Res 43:3403–3407

Bosl GJ, Dmitrovsky E, Reuter VE, Samaniego F, Rodriguez E, Geller NL, Chaganti RSK (1989) Isochromosome of chromosome 12: clinically useful marker for male germ cell tumors. JNCI 81:1874–1878

Castedo SMMJ, de Jong B, Oosterhuis JW, Seruca R, Idenburg VJ, Buist J, Sleijfer DT (1988) i(12p)-Negative testicular germ cell tumors: a different group? Cancer Genet Cytogenet 35:171–178

Castedo SMMJ, de Jong B, Oosterhuis JW, Seruca R, te Meerman GJ, Dam A, Koops HS (1989a) Cytogenetic analysis of ten human seminomas. Cancer Res 49:439–443

Castedo SMMJ, de Jong B, Oosterhuis JW, Idenburg VJS, Seruca R, Buist J, te Meerman GJ, Koops HS, Sleijfer DT (1989b) Chromosomal changes in mature residual teratomas following polychemotherapy. Cancer Res 49:672–676

Castedo SMMJ, de Jong B, Oosterhuis JW, Seruca R, Idenburg VJS, Dam, A, te Meerman GJ, Koops HS, Sleijfer DT (1989c) Chromosomal changes in human primary testicular nonseminomatous germ cell tumors. Cancer Res 49:5696–5701

Chaganti RSK, Ladanyi M, Samaniego F, Reuter V, Jhanwar S, Bosl GJ (1989) Malignant hematopoietic differentiation of a germ cell tumor. Genes Chromosomes Cancer 1:83–89

Dmitrovsky E, Murty VVVS, Moy D, Miller WH Jr, Nanus D, Albino AP, Samaniego F, Bosl G, Chaganti RSK (1990) Isochromosome 12p in nonseminoma cell lines: karyologic amplification of c-Ki-*ras*$_2$ without point-mutational activation. Oncogene 5:543–548

Greco FA, Vaughn WK, Hainsworth JD (1986) Advanced poorly differentiated carcinoma of unknown primary site: recognition of a treatable syndrome. Ann Intern Med 104:547–553

Heim S, Mitelman F (1987) Cancer Cytogenetics. Liss, New York, pp 227–261

Ladanyi M, Samaniego F, Reuter VE, Motzer RJ, Jhanwar SC, Bosl GJ, Chaganti RSK (1990) Cytogenetic and immunohistochemical evidence for the germ cell origin of a subset of acute leukemias associated with mediastinal germ cell tumors. JNCI 82:221–227

Mulder MP, Keijzer W, Splinter TAW, Bos JL (1989) Heterogeneity of *ras* gene activation in human seminoma (Abstr). Satellite Meeting of the 11th International Congress of the International Society of Development Biologists, Groningen

Murty VVVS, Dmitrovsky E, Bosl GJ, Chaganti RSK (1990) Nonrandom chromosome abnormalities in testicular and ovarian germ cell tumor cell lines. Cancer Genet Cytogenet 50:67–73

Samaniego F, Rodriguez E, Houldsworth J, Murty VVVS, Ladanyi M, Lele KP, Chen Q, Dmitrovsky E, Geller NL, Reuter V, Jhanwar SC, Bosl GJ, Chaganti RSK (1990) Cytogenetic and molecular analysis of human male germ cell tumors: chromosome 12 abnormalities and gene amplification. Genes Chromosomes Cancer 1:289–300

Wang L, Vass W, Gao C, Chang KSS (1987) Amplification and enhanced expression of the c-Ki-*ras*$_2$ protooncogene in human embryonal carcinomas. Cancer Res 47:4192–4198

Watmore AE, Potter AM, Sokol RJ, Wood JK (1985) Value of cytogenetic studies in prediction of acute phase CML. Cancer Genet Cytogenet 14:293–301

# *Frequent Occurrence of Activated* ras *Oncogenes in Seminomas but Not in Nonseminomatous Germ Cell Tumors*

M.P. Mulder,[1] W. Keijzer,[1] T.A.W. Splinter,[2] and J.L. Bos[3]

[1] Department of Cell Biology and Genetics, Erasmus University, P.O. Box 1738, 3000 DR, Rotterdam, The Netherlands
[2] Department of Medical Oncology, University Hospital Dijkzigt, Rotterdam, The Netherlands
[3] Laboratory for Molecular Carcinogenesis, Sylvius Laboratories, Leiden, The Netherlands

## Introduction

Little is known about the genetic basis of the development of human testicular germ cell tumors (TGCTs). A common chromosomal abnormality, isochromosome 12p, has been demonstrated in seminomas and nonseminomatous TGCTs (Atkin and Baker 1988), but it is not known whether any particular gene is specifically affected as a result of this aberration. Also, activation of oncogenes has not yet been described for germ cell tumors.

Activated *ras* oncogenes have been identified in a wide variety of human malignancies, the activation being the result of a point mutation at codons 12, 13, or 61 of the Ha- Ki- or N-*ras* gene (for reviews see Barbacid 1987; Bos 1988). The resulting amino acid substitutions cause a decrease in the intrinsic GTPase activity of the p21 *ras* proteins, which would thereby presumably be changed into a constitutively activated form. We have studied the occurrence of *ras* gene mutations in TGCTs. Detailed results of our investigations on seminomas have recently been published (Mulder et al. 1989). Here we briefly review these results and present some preliminary data on nonseminomatous TGCTs.

The presence of activated mutant *ras* genes in DNA extracted from tumor tissue was investigated with specific oligonucleotide hybridization after selective amplification of *ras* gene sequences with the polymerase chain reaction (PCR) as described previously (Mulder et al. 1989; Verlaan-de Vries et al. 1986).

## Seminomas

Seminomas characteristically contain a high proportion of nonneoplastic stromal cells and lymphocytes. Consequently when DNA extracted from seminoma tissue is tested in the PCR/oligonucleotide hybridization assay, a

Recent Results in Cancer Research, Vol. 123

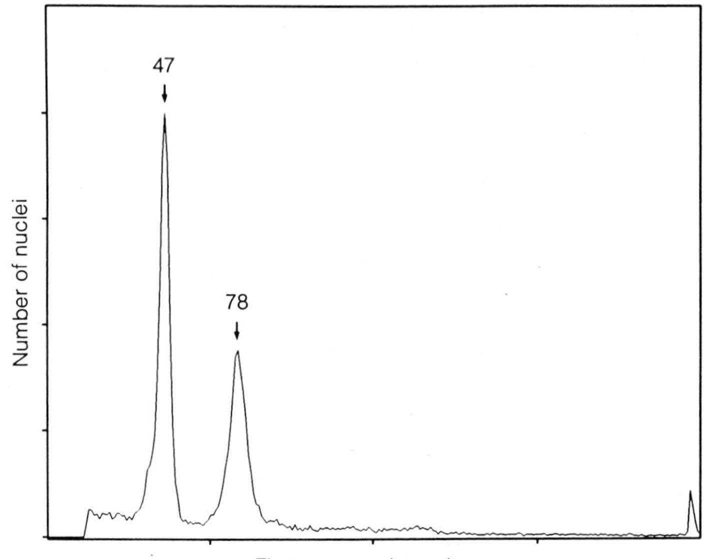


**Fig. 1.** Fluorescence histogram of propidium iodide stained nuclei of tumor HT21. The majority of the nuclei reveal a fluorescence intensity expected for cells with a diploid DNA content (peak at channel No. 47). The fluorescence of the aneuploid nuclei shows a peak value at 1.66 times the diploid value (channel No. 78). Fluorescence intensity was related to DNA content by including chicken red blood cells in a fraction of the sample as an internal standard. (From Mulder et al. 1989)

positive signal for a mutant *ras* gene might be diluted to an undetectable level if the tumor cells constitute only a minor population in the tissue and in addition are heterozygous for the mutation. Therefore isolated tumor cell nuclei were used for the mutation analysis (Mulder et al. 1989). Suspensions of cell nuclei prepared from tissue samples were stained with the DNA-specific fluorochrome propidium iodide and analyzed by flow cytometry. A representative DNA histogram is shown in Fig. 1. In all 14 seminomas investigated, a distinct population of nuclei with an aneuploid DNA content was found. These presumptive tumor cell nuclei were selectively collected by flow sorting. The modal DNA value of the aneuploid population varied between 1.46 and 2.63 (mean, 1.72) times the diploid DNA value for different tumors. In some tumor samples the fraction of aneuploid nuclei was only about 5%, indicating the necessity of the selection procedure.

Extracted DNA from both unfractionated tissue and isolated populations of aneuploid or diploid cell nuclei was analyzed for the presence of mutations at specific positions of the *ras* genes. As illustrated in Fig. 2, a positive signal for a mutant gene is stronger in DNA from isolated aneuploid nuclei than in DNA from an unfractionated population. In addition the mutation is not detected in the diploid population, indicating the absence of diploid host and/or tumor cells carrying *ras* mutations. Apparently the separation procedure is effective.

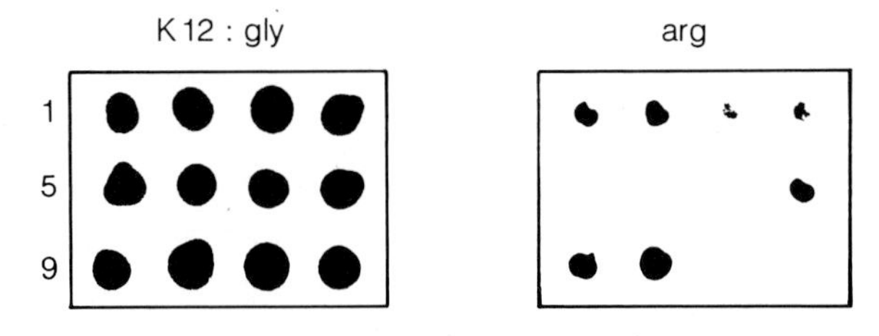


**Fig. 2.** Dot-blot hybridization of DNA extracted from seminoma tissue or isolated nuclei. DNA sequences spanning codon 12 of the Ki-*ras* gene were amplified in vitro and hybridized to synthetic oligonucleotide probes specific for the wild-type sequence (*gly*) or a codon 12 mutant (*arg*). Sources of spotted DNAs: *1–4*, unfractionated samples of tumor HT85; *5–7*, diploid nuclei isolated from samples 2, 3, and 4; *8–10*, aneuploid nuclei isolated from samples 2, 3, and 4; *11,12*, control tissue. (From Mulder et al. 1989)

**Table 1.** *ras* gene mutations in seminoma

| Tumor | Age[a] | Clinical stage[b] | DNA index[c] | *ras* mutation[d] |
|---|---|---|---|---|
| HT1 | 29 | IIC | 1.46 | N-61, arg |
| HT14 | 39 | IIC | 1.68 | N-12, asp |
| HT21 | 30 | IIC | 1.66 | N-61, arg |
| HT27 | 33 | I | 1.50 | N-61, arg |
| HT31 | 26 | I | 1.64 | Ki-12, val |
| HT85 | 24 | I | 1.44 | Ki-12, arg |
| HT3 | 27 | I | 2.63 | – |
| HT5 | 38 | IIB | 1.55 | – |
| HT45 | 52 | I | 1.51 | – |
| HT48 | 29 | IVH | 1.52 | – |
| HT51 | 39 | IIIN1 | 1.94 | – |
| HT74 | 37 | I | 2.33 | – |
| HT86 | 38 | I | 1.55 | – |
| HT95 | 41 | I | 1.65 | – |

[a] Age of patient (years) at presentation.
[b] Staging of the disease at presentation according to the Royal Marsden Hospital classification (Peckham et al., 1979).
[c] DNA content of aneuploid nuclei relative to the DNA content of diploid nuclei.
[d] Position of mutation and amino acid substitution as determined by oligonucleotide hybridization.

The results of the hybridization assays are summarized in Table 1. Mutations were found at codons 12 or 61 of either the Ki-*ras* or the N-*ras* gene. No Ha-*ras* mutations were detected. In total 6 out of 14 tumors, i.e., 43%, carried a mutation. There appears to be no correlation of the presence of a mutant *ras* gene with the clinical stage of the tumor (Table 1). In

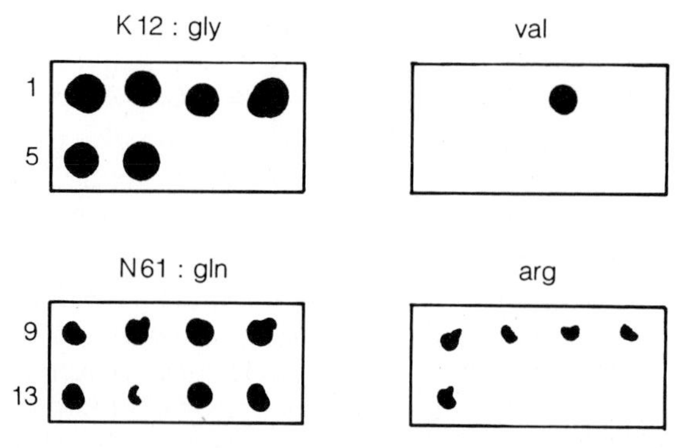


**Fig. 3.** Dot-blot hybridization of DNA extracted from aneuploid nuclei isolated from small fragments of seminoma tissue. DNA sequences spanning codon 12 of the Ki-*ras* gene or codon 61 of the N-*ras* gene were amplified and hybridized to oligonucleotide probes specific for the wild type (K12,gly and N61,gln) or mutant (K12,val and N61,arg) sequences. Sources of spotted DNAs: *1,2*, control tissue; *3–6*, aneuploid nuclei isolated from four small fragments of tumor HT31; *9–13*, aneuploid nuclei isolated from five small fragments of tumor HT1; *14–16*, control tissue. (From Mulder et al. 1989)

addition, microscopic examination of tumor sections revealed no differences between tumors with and without a mutation.

## Tumor Heterogeneity

Figure 2 shows that the signal for the mutant *ras* gene is weaker than the signal for the wild-type allele in the same DNA sample, even in DNA from isolated aneuploid nuclei. This was found in all six tumors in which a mutation was detected, suggesting that either the aneuploid tumor cells carry an excess of wild-type gene copies or the mutation is present in only a fraction of the tumor cell population. Support for the latter hypothesis was obtained by analyzing the occurrence of the *ras* mutation in various small fragments (10–20 $mm^3$) of the tumors that had scored positively. An example of the results is presented in Fig. 3. In three tumors (HT14, HT21, and HT31) both positive and negative areas were detected. In the other three positive tumors no such heterogeneity was found, but here the signal for the mutant *ras* gene was weaker than the wild-type signal in all the fragments analyzed (HT1 in Fig. 3).

We conclude that in some and possibly most seminomas carrying a mutant *ras* gene the mutation is present in only a fraction of the tumor cell population. Yet, histological examination of sections and DNA flow cytometry did not reveal clonal heterogeneity. Within a tumor the DNA content of the aneuploid nuclei was always the same in tumor fragments with and without a mutant *ras* gene.

### Nonseminomas

Recently we investigated the occurrence of *ras* gene mutations in a number of nonseminomatous TGCTs. Flow cytometric analysis of nuclear DNA content revealed that in these tumors the fraction of diploid cells is generally lower than in seminomas. Therefore in most cases the assay for the detection of *ras* mutations could be carried out without prior flow sorting of aneuploid nuclei. Twelve tumors were investigated: one pure embryonal carcinoma, nine malignant nonseminomatous TGCTs with various differentiated components (two of these contained a prominent trophoblastic component and two others a prominent yolk sac component), and two mature teratomas. In none of these tumors was a mutant *ras* gene detected.

### Discussion

We have detected a mutated Ki- or N-*ras* gene in about 40% of the investigated seminomas. This incidence is in the same range as reported for acute myeloid leukemia and for carcinomas of the colon, lung, and thyroid (Bos 1988). A significantly higher incidence (over 90%) has only been detected in pancreatic carcinoma (Almoguera et al. 1988), while in many other tumor types the incidence is much lower (0%–10%). In seminomas the presence or absence of a mutant *ras* gene is not correlated with specific clinical or histopathological features. Seminomas are characteristically very uniform. Also in those seminomas that carried a mutation only in a subpopulation of the tumor cells, no heterogeneity was apparent within the tumor by other criteria. Heterogeneity within a tumor with respect to the presence of a *ras* mutation has recently been described for several cases of acute myeloid and lymphoblastic leukemia (Shen et al. 1987; Toksoz et al. 1987; Farr et al. 1988; Neri et al. 1988) and for a hepatoma (Takada and Koike 1989).

Although the heterogeneity within the seminomas can be interpreted in various ways, we prefer the hypothesis that the *ras* mutation has been acquired by a subpopulation of the tumor cells in the course of tumor development (Mulder et al. 1989). This event probably took place after the establishment of a clonal population of aneuploid cells since within a seminoma the aneuploid cells with and without a mutation had the same nuclear DNA content. The primary initiating events may have resulted in a clonal population of slowly growing abnormal cells. Fully transformed, faster growing cells might then arise by different secondary events, one of which could be a *ras* mutation. If such secondary events occurred on two or more separate occasions, then a heterogeneous seminoma would be the result. A similar development has been suggested for a number of cases of acute myeloid leukemia (Farr et al. 1988; Yunis et al. 1989). We conclude that the *ras* mutation does play a role in the development of seminoma but probably is not the initial genetic event.

We have so far not been able to detect mutated *ras* genes in nonseminomatous TGCTs. Thus, although seminomas and nonseminomatous TGCTs probably both originate from primordial germ cells, *ras* gene activation appears to be associated with only one developmental lineage. Our data do not allow a choice between a pathogenetic model in which seminomas and nonseminomatous TGCTs develop separately (Pierce and Abell 1970) or a model in which nonseminomatous TGCTs are thought to pass through a seminoma stage (Friedmann 1951; Oosterhuis et al. 1989). However, the evolution of these tumor types is different with respect to acquisition of *ras* mutations.

Recent data indicate that carcinoma *in situ* (CIS) of the testis may be a common precursor lesion for all types of TGCTs with the possible exception of spermatocytic seminoma and infantile yolk sac tumor (Skakkebaek and Berthelsen 1981; Skakkebaek et al. 1987; Koide et al. 1987). Our data on the occurrence of activated *ras* genes suggest that this gene mutation is not involved in the development of CIS, but may play a specific role in the progression of CIS to seminoma.

## Conclusion

Forty percent of seminomas carry a mutation at codon 12 or 61 of either the Ki- or N-*ras* gene. No correlation was found between histopathological or clinical features and the presence of an activated *ras* oncogene. In some seminomas the mutation was present in only a fraction of the tumor cell population, suggesting tumor heterogeneity. Yet, DNA flow cytometry and histological examination failed to show distinct tumor cell subpopulations within a seminoma. No *ras* gene mutations were found in nonseminomatous TGCTs. These data suggest that oncogenic activation of a *ras* gene is not a primary initiating event in the development of TGCTs but may be acquired in the course of progression of a seminoma.

*Acknowledgements.* We are grateful to J.C.M. Langeveld and M. Burghouwt for technical assistance in working up the tumor samples, to A. Verkerk for flow cytometry and sorting, to A.J.M. Boot and M. Verlaan-de Vries for PCR/oligonucleotide hybridization assays, to Drs. M.E.F. Prins and F.W. ten Kate for histopathological judgement of tumor samples, to P.C. van Sluijsdam and R.J. Boucke for typing the manuscript, and to Dr. D. Bootsma for fruitful discussions.

## *References*

Almoguera C, Shibata D, Forrester K, Martin J, Arnheim N, Perucho M (1988) Most human carcinomas of the exocrine pancreas contain mutant c-K-*ras* genes. Cell 53:549–554

Atkin NB, Baker MC (1988) Small metacentric marker chromosomes, particularly isochromosomes, in cancer. Hum Genet 79:96–102

Barbacid M (1987) *ras* genes. Annu Rev Biochem 56:779–827

Bos JL (1988) The *ras*-gene family and human carcinogenesis. Mutat Res 195:255–271

Farr CJ, Saiki RK, Erlich HA, McCormick F, Marshall CJ (1988) Analysis of *ras* gene mutations in acute myeloid leukemia by polymerase chain reaction and oligonucleotide probes. Proc Natl Acad Sci USA 85:1629–1633

Friedman NB (1951) The comparative morphogenesis of extragenital and gonadal teratoid tumors. Cancer 4:265–276

Koide O, Iwai S, Baba K, Iri H (1987) Identification of testicular atypical germ cells by an immunohistochemical technique for placental alkaline phosphatase. Cancer 60:1325–1330

Mulder MP, Keijzer W, Verkerk A, Boot AJM, Prins MEF, Splinter TAW, Bos JL (1989) Activated *ras* genes in human seminoma: evidence for tumor heterogeneity. Oncogene 4:1345–1351

Neri A, Knowles DM, Greco A, McCormick F, Dalla-Favera R (1988) Analysis of *ras* oncogene mutations in human lymphoid malignancies. Proc Natl Acad Sci USA 85:9268–9272

Oosterhuis JW, Castedo SMMJ, de Jong B, Cornelisse CJ, Dam A, Sleijfer DT, Schraffordt Koops H (1989) Ploidy of primary germ cell tumors of the testis. Pathogenetic and clinical relevance. Lab Invest 60:14–21

Peckham MJ, Barret A, McElwain TJ, Hendry WF (1979) Combined management of malignant teratoma of the testis. Lancet 2:267–270

Pierce GB, Abell MR (1970) Embryonal carcinoma of the testis. Pathol Annu 5:27–60

Shen WPV, Aldrich TH, Venta-Perez G, Franza BR, Furth ME (1987) Expression of normal and mutant *ras* proteins in human acute leukemia. Oncogene 1:157–165

Skakkebæk NE, Berthelsen JG (1981) Carcinoma-in-situ of the testis and invasive growth of different types of germ cell tumours. A revised germ cell theory. Int J Androl [Suppl] 4:26–33

Skakkebæk NE, Berthelsen JG, Giwercman A, Müller J (1987) Carcinoma-in-situ of the testis: possible origin from gonocytes and precursor of all types of germ cell tumors except spermatocytoma. Int J Androl 10:19–28

Takada S, Koike K (1989) Activated N-*ras* gene was found in human hepatoma tissue but only in a small fraction of the tumor cells. Oncogene 4:189–193

Toksoz D, Farr CJ, Marshall CJ (1987) *ras* gene activation in a minor proportion of the blast population in acute myeloid leukemia. Oncogene 1:409–413

Verlaan-de Vries M, Boogaard ME, van den Elst H, van Boom JH, van der Eb AJ, Bos JL (1986) A dot-blot screening procedure for mutated *ras* oncogenes using synthetic oligodeoxynucleotides. Gene 50:313–320

Yunis JJ, Boot AJM, Mayer MG, Bos JL (1989) Mechanisms of *ras* mutation in myelodysplastic syndrome. Oncogene 4:609–614

# *Differential Activation of Homeobox Genes by Retinoic Acid in Human Embryonal Carcinoma Cells*

L. Bottero,[1] A. Simeone,[2] L. Arcioni,[3] D. Acampora,[2]
P.W. Andrews,[4] E. Boncinelli,[2] and F. Mavilio[3]

[1] Department of Hematology-Oncology, Istituto Superiore di Sanitá, Roma, Italy
[2] International Institute of Genetics and Biophysics, CNR, Napoli, Italy
[3] Istituto Scientifico H.S. Raffaele, Milano, Italy
[4] The Wistar Institute, Philadelphia, PA 19104, USA

## Introduction

Morphogens, i.e., endogenous, diffusible molecules which induce pathways of differentiation through a gradient of concentration, play a key role in vertebrate development. Retinoic acid (RA) is a natural morphogen in chicken development, where it specifies the limb anteroposterior (AP) axis (Tickle et al. 1982; Thaller and Eichele 1987, 1988; Brockes 1989), and possibly in frogs, where alteration of intraembryonic RA levels dramatically affects the AP polarity of the developing CNS (Durston et al., 1989). The identification of three nuclear RA receptors in mouse and man (Petkovich et al. 1987; Giguere et al. 1987; Brand et al. 1988; Zelent et al. 1989) supports the concept that the morphogens act as intracellular signal molecules with direct gene control functions. Although the genes involved in transducing signals provided by morphogens are still unknown, homeobox genes, which specify positional information in *Drosophila* and possibly in vertebrate embryogenesis (Gehring 1987; Ingham 1988; Holland and Hogan 1988), are among the suggested candidates (Brockes 1989; de Robertis et al. 1989).

Four different clusters of *Antennapedia*-like homeobox genes have been identified in the human genome, namely *HOX1, 2, 3*, and *4*, located on chromosomes 7, 17, 12, and 2, respectively (Bucan et al. 1986; Rabin et al. 1986; Cannizzaro et al. 1987). All clusters contain at least eight different genes, organized in a homologous linear manner whereby genes sharing the same position in each cluster also share the highest homology in the homeobox sequence (Acampora et al. 1989) (Fig. 1). The same, conserved organization has been observed in the mouse genome (Duboule and Dollè 1989). Previous studies showed that homeobox genes are expressed and developmentally regulated in human (Simeone et al. 1986, 1987; Mavilio et al. 1986) and mouse embryos (reviewed in Holland and Hogan 1988). Virtually all of them are expressed in the developing CNS. In particular, multiple genes of each cluster are expressed along the CNS AP axis in the A to P direction, following a gradient which reflects their relative positions within the cluster in the 3′ to 5′ orientation (Graham et al. 1989; Wilkinson et al. 1989; Giampaolo et al. 1989; Duboule and Dollè 1989). A similar observation was made with the mouse *Hox5* (= human *HOX4*) genes, which are expressed along the AP axis of the developing limb (Dollè et al. 1989a). These intriguing colinearities might be the basis of a specific role of mammalian homeobox genes in specifying positional information, possibly by transducing signals provided by morphogen gradients (Brockes 1989; de Robertis et al. 1989).

We have previously reported that RA specifically activates the expression of all tested *HOX1, 2, 3*, and *4* genes in the embryonal carcinoma (EC) cell line NTera 2, clone D1 (NT2/D1), thus providing a convenient in vitro model to study their regulation (Mavilio et al. 1988). NT2/D1 cells show a phenotypic pattern of multipotent embryonic stem cells, and are induced to differentiate by RA into a variety of cell types, including neurons (Andrews 1984). In an attempt to gain more information on *HOX* gene regulation by RA, we studied the effect of both concentration and exposure time to RA on *HOX1, 2*, and *4* gene expression in differentiating NT2/D1 cells.

## Activation of *HOX* Genes by RA in a Concentration-Dependent Fashion

NT2/D1 cells were cultured in medium containing RA at concentrations varying from $10^{-8}$ to $10^{-5}$ $M$ for 14 days and the extent of differentiation monitored by the proportion of cells positive to glycolipid surface markers characteristically expressed in either stem or differentiated EC cells (Andrews 1987). An RA concentration as low as $5 \times 10^{-8}$ $M$ is sufficient to cause the disappearance of 50% of the stem cells positive to SSEA-3, SSEA-4, and TRA1–60, and almost plateau expression of differentiation antigens A2B5 and ME311 in 14-day cultures (not shown). NT2/D1 cells thus respond to RA at concentrations three orders of magnitude lower than those originally reported as optimal for complete differentiation (Andrews 1984).

Total RNA was extracted from 14-day cultures, poly $(A)^+$ selected, run on a Northern blot assay, and probed for expression of the nine genes of the *HOX2* cluster, two genes from the *HOX1* cluster (*HOX1C* and *1D*), and two genes from the *HOX4* cluster (*HOX4A* and *4B*). The number and size of transcripts from all *HOX* genes in EC cells are constant in all tested conditions, and faithfully reproduce the patterns observed in human embryonic CNS (Simeone et al. 1986, 1987; Mavilio et al. 1986; Giampaolo et al. 1989). In the *HOX2* cluster, an RA concentration in the order of $10^{-8}$ $M$ is sufficient to induce peak accumulation of transcripts from genes located in the 3′ part of the cluster, i.e., *HOX2F, 2H*, and *2I*, and no significant variation was observed in the presence of increasing RA concentration. Conversely, transcripts from the 5′ genes *HOX2E, 2D, 2C, 2B*, and *2A* are accumulated at significant levels in the presence of at least $5 \times 10^{-6}$ $M$ RA, whereas much lower RNA levels are detected in cells exposed to lower RA concentrations. Expression of *HOX2D, 2A, 2H*, and *2I* is shown as an example in Fig. 2. These data indicate that homeobox genes on the *HOX2* cluster are activated at maximum levels by RA concentrations that are two orders of magnitude lower than those required to activate fully the most 5′ ones. The border between the two groups is apparently at the level of *HOX2A*. We therefore analyzed the expression pattern of *HOX1C* and *1D*, homologous on the *HOX1* cluster to *HOX2A* and *2F* respectively (see Fig. 1). Expression of *HOX1D* was detectable without significant variation in cells treated with $10^{-8}$ to $10^{-5}$ *M* RA, whereas *HOX1C* transcripts were accumulated at maximal levels starting from $10^{-6}$ *M* (Fig. 3). These data indicate that the mechanism underlying the concentration-dependent differential activation of *HOX* genes by RA is conserved in at least two clusters, i.e., *HOX1* and *HOX2*.

We analyzed the expression pattern of two genes from the *HOX4* cluster, i.e., *HOX4B*, homologous to *HOX2F* and *HOX1D* and *HOX4A*, homologous to *HOX2G* (Fig. 1). Transcripts from both genes were accumulated at peak levels at the lowest RA concentration, in accordance with their location in the 3′ region of the cluster (Fig. 3). However, *HOX4B*, which we have previously shown to be expressed in human embryogenesis according to a complex spatially restricted and stage-specific pattern (Mavilio, et al. 1986), showed a peculiar expression pattern. On Northern analysis, a 4.2-kb transcript is accumulated in EC cells after exposure to the $10^{-8}$ *M* RA, whereas 5.4-kb and 2.8-kb transcripts are increasingly accumulated starting from $10^{-6}$ *M* RA (Fig. 3). We previously reported that the anterior border of expression of the 4.2-kb transcript in 6- to 7-week-old human embryonic CNS is in the hindbrain region, whereas 5.4- and 2.8-kb transcripts are expressed more

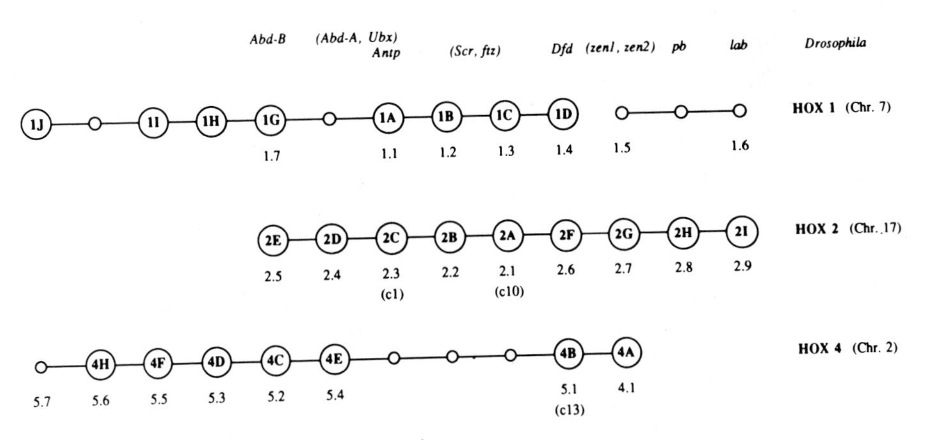


**Fig. 1.** Schematic representation of the human *HOX1, HOX2*, and *HOX4* loci. *HOX* genes are indicated by *circles. Small circles* indicate genes absent or not found yet in the human genome. Corresponding mouse genes are indicated *below each circle, Drosophila* genes are indicated in *italics*. Designations *between parentheses* refer to laboratory names of previously published cDNA clones

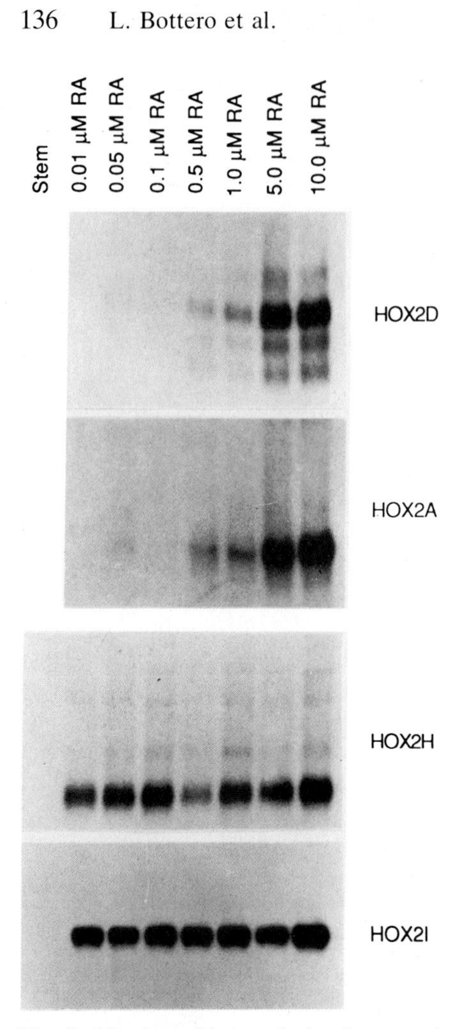


**Fig. 2.** Northern blot analysis of expression of *HOX2* genes in poly(A)$^+$ RNA from the human embryonal carcinoma cell line NT2/D1 before (control) and after 14 days induction with $10^{-8}$–$10^{-5}$ *M* retinoic acid (*RA*)

caudally in the spinal cord, as well as in the limb bud (Mavilio et al. 1986). These data indicate that different RA concentrations direct alternative expression patterns from *HOX4B* in EC cells in a manner reminiscent of those observed in vivo in developing embyros.

## Sequential Activation of *HOX2* Genes by RA in EC Cells

NT2/D1 cells were treated with $10^{-5}$ *M* RA and analyzed for the expression of single *HOX2* genes at regular intervals by RNase protection assay.

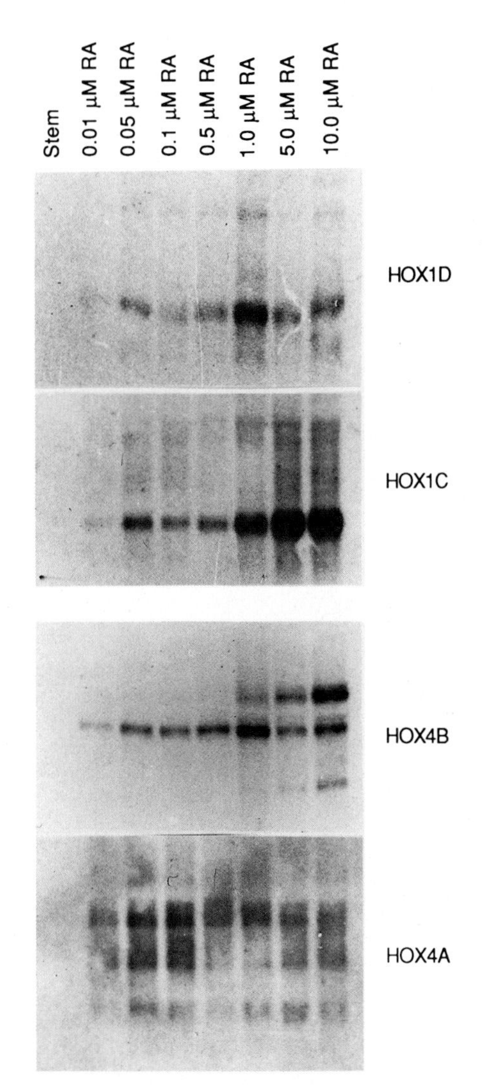


**Fig. 3.** Northern blot analysis of expression of *HOX1* and *HOX4* genes in NT2/D1 cells before (*control*) and after 14 days induction with $10^{-8}$–$10^{-5}$ *M* retinoic acid (*RA*)

Transcripts from the most 3′ *HOX2H* and *2I* genes are rapidly and steadily accumulated in RA-treated cells, starting from very low levels (undetectable by Northern assay) in untreated cells and reaching peak levels at 24 h of RA exposure. Activation of the other *HOX2* genes proceeds sequentially from *HOX2G* to *HOX2E* between 1 and 7 days of RA induction. No transcript from these genes is detected in untreated cells. Figure 4 summarizes the induction kinetics of RA-induced *HOX2* gene activation, which proceeds along the cluster in the 3′ to 5′ direction, according to a distinct pattern.

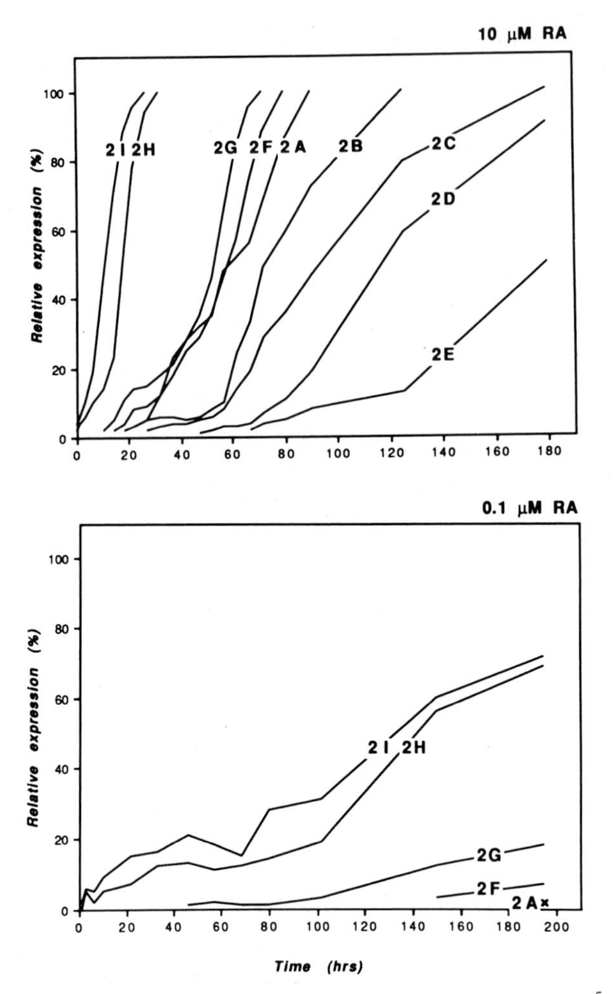


**Fig. 4.** Kinetics of *HOX2* gene induction by *RA* at $10^{-5}$ *M* (*upper panel*) or $10^{-7}$ *M* (*lower panel*) concentration in NT2/D1 cells. Graphs are obtained by densitometric scanning of RNAse protection autoradiographies. Expression values are in arbitrary units, where 100 represents plateau RNA accumulation levels. Time is in hours

Continuous presence of RA is required throughout the induction process in order to activate the most 5′ genes. In fact, we already showed that 1–4 days RA exposure are not sufficient to induce stable accumulation of transcripts from *HOX2A, 2B, 2C, 2D*, and *2E* genes in differentiated EC cells (Mavilio et al. 1988).

The time course of *HOX2* gene activation was also analyzed in cells induced by $10^{-7}$ *M* RA (Fig. 4). Genes in the 3′ block are activated according to much slower kinetics than with the $10^{-5}$ *M* RA induction, even though comparable peak levels of RNA accumulation are eventually reached. Transcripts from genes in the 5′ block were instead undetectable up to 8 days from RA exposure, confirming their requirement for higher RA levels for activation.

## Expression of RA Receptors in EC Cells

We analyzed the expression of the α, β, and γ RA receptors (*hRAR*-α, *hRAR*-β, and *hRAR*-γ) in EC cells before and after RA induction, in order to study the correlation between expression of specific receptors and *HOX* gene activation. Northern blot analysis showed transcripts for *hRAR*-α and *hRAR*-γ in both uninduced and induced EC cells whereas *hRAR*-β transcripts were abundantly accumulated only in RA-treated cells, at levels that do not substantially change with RA concentrations (Fig. 5). Appearance of *hRAR*-β transcripts after induction closely follows that of the early responders *HOX2I* and *2H* (not shown). These data, and the fact that *HOX2* genes can also be induced in the absence of the novo protein synthesis, and therefore of functional β-receptors, indicate that *hRAR*-β may be not necessary in mediating the activation process, at least in EC cells. *RAR*-β is abundantly and differentially expressed in mouse embryonic tissues (Dollè et al. 1989b), and its inducibility and higher affinity for the ligand still suggest that it could play a key role in the fine tuning of homeobox gene activation in response to RA in vivo.

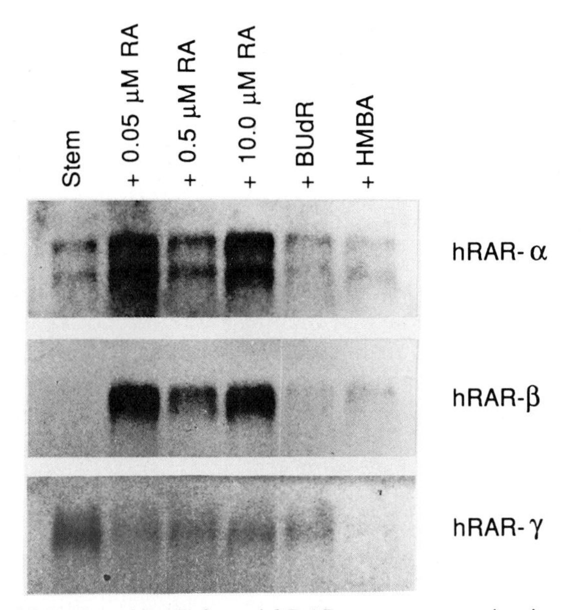


**Fig. 5.** Northern blot analysis of *hRAR*-α, *hRAR*-β, and *hRAR*-γ gene expression in NT2/D1 cells before (control) and after induction with RA at three different concentrations, hexamethylene bisacetamide (*HMBA*), or bromodeoxyuridine (*BUdR*)

## Discussion

We have shown that human *HOX2* genes are differentially activated in EC cells by RA in a concentration-dependent fashion and in a sequential order which is colinear with their 3′ to 5′ arrangement in the cluster. The same polarity is observed in the expression pattern of *HOX2* genes along the AP axis of the developing CNS, where 3′ genes are expressed more rostrally in myelencephalon and 5′ genes more caudally in the spinal cord (Giampaolo et al. 1989). A more detailed analysis is available in the CNS of mouse early embryos, where the anterior boundary of expression of *HOX2.8* (*HOX2H*), *2.7* (=*2G*), and *2.6* (=*2F*) are at the level of the 3rd, 5th, and 7th rhombomere respectively in the hindbrain, whereas genes from *Hox 2.1* (=*HOX2A*) upstream are all expressed progressively more caudally in the spinal cord (Wilkinson et al. 1989). The present data, although obtained in EC cells in vitro, suggest that a putative RA concentration gradient operating in vivo might at least in part underlie a differential activation of *HOX2* genes consistent with the spatially restricted expression pattern observed in vivo. Concentration levels and/or exposure time to the morphogen of cells in different regions of the embryo could be critical factors in determining what genes are to be locally activated. The differential activation pattern of *HOX2* genes could in turn provide positional information required for further developmental decisions. In this regard, it is suggestive that *HOX2A*, which marks the transition from hindbrain to spinal cord in vivo, also marks the border between genes requiring high or low levels of RA in order to be activated in vitro.

Evidence supporting a role for RA in AP polarity determination is provided by experiments in *Xenopus* embryos, where alteration of intraembryonic RA levels leads to pronounced caudalization of the whole CNS (Durston et al. 1989). These embryos lack of the forebrain and midbrain as well as the sense organs induced by or formed from the forebrain (eyes and nasal pits). Formation of more posterior neural structures and ears appears unaffected. It would be crucial to know if these RA-induced transformations are associated with overexpression and/or ectopic expression of homeobox genes. It is suggestive that the most anterior expression boundaries of *HOX2* genes observed in mammalian embryos coincide with the border between affected and nonaffected structures in *Xenopus* embryos. Altogether, these and our studies provide suggestive evidence for an important role of homeobox genes in transducing information provided by RA into the AP pattern specification in vertebrate development.

## Materials and Methods

The cloned, human EC cell line NT2/D1 (Andrews 1984) was maintained in high-glucose Dulbecco's modified Eagle's medium supplemented with 10%

fetal calf serum in a humidified atmosphere of 5% $CO_2$ in air. Differentiation was induced by seeding cells at $5 \times 10^5$ per 75-$cm^2$ flask in $10^{-8}$ to $10^{-5}$ *M* RA (all-*trans*, Sigma Chemical Co.), $3 \times 10^{-3}$ *M* hexamethylene bisacetamide (HMBA), or $3 \times 10^{-6}$ *N* bromouracil deoxyribose (BUdR). Cells were refed every 4 days with medium containing the inducers and analyzed after 14 days by indirect immunofluorescence, as previously described (Andrews 1984, 1987).

Total RNA was extracted by the guanidine-isothiocyanate technique (Chirgwin et al. 1979), selected for poly$(A)^+$ by one passage on oligo (dT) cellulose columns according to standard procedure, run in 2- to 3-µg aliquots on 1.0% agarose/formaldehyde gel, transferred by Northern capillary blot (Thomas 1980) onto nylon membranes (Hybond-N, Amersham), and hybridized to $10^7$ cpm DNA probe labeled by random priming to a specific activity of $10^9$ dpm/µg. Filters were washed under stringent conditions (i.e., 0.1 × SSC, 0.2% SDS, at 65°C for 60 min) and exposed to Kodak XAR-5 films at −70°C for 1–3 days. Filters were stripped and reprobed up to six times as previously described (Mavilio et al. 1988). *HOX2* probes were: a 3.0-kb *Sma*1 genomic fragment of *HOX2E*, a 1.5-kb *Eco*R1/*Bgl*II fragment of *HOX2D*, a 1.3-kb cDNA clone (*HHO.c1.95*) from *HOX2C* (Simeone et al. 1987), a 2.0-kb *Eco*R1 fragment of *HOX2B*, a 1.2-kb cDNA clone (*HHO. c10*) from *HOX2A* (Simeone et al. 1986), a 0.3-kb *Hin*dIII fragment of *HOX2F*, a 0.8-kb *Bbl*II/*Eco*R1 fragment of *HOX2G*, a 2.8-kb *Eco*R1 fragment of *HOX2H*, and a 0.6-kb *Pvu*II fragment of *HOX2I*. *HOX1* probes were a 1.5-kb *Bgl*2 genomic fragment of *HOX1C* and a 1.1-kb *Hin*dIII genomic fragment of *HOX1D*. *HOX4* probes were a 1.3-kb cDNA clone (*HHO.c13*) from *HOX4B* (Mavilio et al. 1986) and a 0.65-kb *Bgl*2/*Eco*R1 genomic fragment of *HOX4A*.

RNase protection experiments were carried out essentially as described by Melton et al. (1984). Partial 3′ noncoding sequences from *HOX2* genes were subcloned in pGEM vectors (Promega Biotec.). A *Sma*1/*Eco*R1 fragment from hRAR-α an *Eco*R1/*Bam*HI fragment from hRAR-β cDNAs (Petkovich et al. 1987; Brand et al. 1988) were subcloned in pBluescript vectors (Stratagene). Antisense strand RNA probes were synthesized with Sp6 or T7 polymerase and hybridized to 20 µg RNA at appropriate temperatures. RNase digestion and electrophoresis on 7% urea/polyacrylamide gels were carried out as described (Melton et al. 1984).

*Acknowledgements.* We are indebted to Dr. Martin Petkovich and Prof. Pierre Chambon for generously providing *hRAR*-α, *hRAR*-β, and *hRAR*-γ cDNA sequences. This work has been supported by grants from the Associazione Italiana per la Ricerca sul Cancro (AIRC) and the CNR Special Project "Ingegneria Genetica," and by USPHS grant CA29894 from NIH.

***References***

Acampora D, D'Esposito M, Faiella A, Pannese M, Migliaccio E, Morelli F, Stornaiuolo A, Nigro V, Simeone A, Boncinelli E (1989) The human *HOX* gene family. Nucl Acids Res 17:10385–10402

Andrews PW (1984) Retinoic acid induces neuronal differentiation of a cloned human embryonal carcinoma cell line in vitro. Dev Biol 103:285–293

Andrews PW (1987) Human teratocarcinoma stem cells: glycolipid antigen expression and modulation during differentiation. J Cell Biochem 35:321–332

Brand N, Petkovich M, Krust A, Chambon P, de The H, Marchio A, Tiollais P, Dejean A (1988) Identification of a second human retinoic acid receptor. Nature 332:850–853

Brockes JP (1989) Retinoids, homeobox genes and limb morphogenesis. Neuron 2:1285–1294

Bucan M, Yang-Feng T, Colberg-Poley AM, Wolgemuth DJ, Guenet JL, Francke U Lehrach H (1986) Genetic and cytogenetic localization of the homeo box containing genes on mouse chromosome 6 and human chromosome 7. Embo J 5:2899–2905

Cannizzaro LA, Croce CM, Griffin CA, Simeone A, Boncinelli E, Huebner K (1987) Human homeo box-containing genes located at chromosome regions 2q31–2q37 and 12q12–12q13. Am J Hum Genet 41:1–15

Chirgwin JM, Przybyla AE, MacDonald RJ, Rutter WJ (1979) Isolation of biologically active ribonucleic acid from sources enriched in ribonuclease. Biochemistry 18: 5294–5300

De Robertis EM, Oliver G, Wright CVE (1989) Determination of axial polarity in the vertebrate embryo: homeodomain proteins and homeogenetic induction. Cell 57: 189–191

Dollè P, Izpisua-Belmonte JC, Falkenstein H, Renucci A, Duboule D (1989a) Coordinate expression of the murine *HOX-5* complex homeobox-containing genes during limb pattern formation. Nature 342:767–772

Dollè P, Ruberte E, Kastner P, Petkovich M, Stoner CM, Gudas LJ, Chambon P (1989b) Differential expression of genes encoding α,β and γ retinoic acid receptors and CRABP in the developing limbs of the mouse. Nature 342:702–705

Duboule D, Dollè P (1989) The structural and functional organization of the murine *HOX* gene family resembles that of *Drosophila* homeotic genes. Embo J 8: 1497–1505

Durston AJ, Timmermans JPM, Mage WJ, Hendbiks HFJ, de Vries NJ, Heideveld M, Nievwkoop PD (1989) Retinoic acid causes an anteroposterior transformation in the developing central nervous system. Nature 340:140–144

Gehring WJ (1987) Homeo boxes in the study of development. Science 236: 1245–1252

Giampaolo A, Acampora D, Zappavigna V, Pannese M, D'Esposito M, Care' A, Faiella A, Stornaiuolo A, Russo G, Simeone A, Boncinelli E, Peschle C (1989) Differential expression of human *HOX-2* genes along the anterior-posterior axis in embryonic central nervous system. Differentiation 40:191–197

Giguere V, Ong ES, Segui P, Evans RM (1987) Identification of a receptor for the morphogen retinoic acid. Nature 330:624–629

Graham A, Papalopulu N, Krumlauf R (1989) The murine and *Drosophila* homeobox gene complexes have common features of organization and expression. Cell 57: 367–378

Holland PWH, Hogan BLM (1988) Expression of homeo box genes during mouse development: a review. Genes Dev 2:773–782

Ingham PW (1988) The molecular genetics of embryonic pattern formation in *Drosophila*, Nature 335:25–34

Mavilio F, Simeone A, Giampaolo A, Faiella A, Zappavigna V, Acampora D, Poiana G, Russo G, Peschle C, Boncinelli E (1986) Differential and stage-related expression in embryonic tissues of a new human homeobox gene. Nature 324:664–668

Mavilio F, Simeone A, Boncinelli E, Andrews PW (1988) Activation of four homeobox gene clusters in human embryonal carcinoma cells induced to differentiate by retinoic acid. Differentiation 37:73–79

Melton DA, Krieg PA, Rebagliati MA, Maniatis T, Zinn K, Green MR (1984) Efficient in vitro synthesis of biologically active RNA and RNA hybridization probes from plasmid containing a bacteriophage Sp6 promoter. Nucl Acids Res 12:7035–7056

Petkovich M, Brand NJ, Krust A, Chambon P (1987) A human retinoic acid receptor which belongs to the family of nuclear receptors. Nature 330:444–450

Rabin M, Ferguson-Smith A, Hart CP, Ruddle FH (1986) Cognate homeo-box loci mapped on homologous human and mouse chromosomes. Proc Natl Acad Sci USA 83:9104–9108

Simeone A, Mavilio F, Bottero L, Giampaolo A, Russo G, Faiella A, Boncinelli E, Peschle C (1986) A human homeobox gene specifically expressed in spinal cord during embryonic development. Nature 320:763–765

Simeone A, Mavilio F, Acampora D, Giampaolo A, Faiella A, Zappavigna V, D'Esposito M, Pannese M, Russo G, Boncinelli E, Peschle C (1987) Two human homeobox genes, *c1* and *c8*: structure analysis and expression in embryonic development. Proc Natl Acad Sci USA 84:4914–4918

Thaller C, Eichele G (1987) Identification and spatial distribution of retinoids in the developing chick limb bud. Nature 327:625–628

Thaller C, Eichele G (1988) Characterization of retinoid metabolism in the developing chick limb bud. Development 103:473–483

Thomas PS (1980) Hybridization of denatured RNA and small DNA fragments transferred to nitrocellulose. Proc Natl Acad Sci USA 77:5201–5205

Tickle C, Alberts BM, Wolpert L, Lee J (1982) Local application of retinoic acid to the limb bud mimics the action of the polarizing region. Nature 296:564–565

Wilkinson DG, Bhatt S, Cook M, Boncinelli E, Krumlauf R (1989) Segmental expression of *HOX-2* homeobox-containing genes in the developing mouse hindbrain. Nature 341:405–409

Zelent A, Krust A, Petkovich M, Kastner P, Chambon P (1989) Cloning of murine α and β retinoic acid receptors and a novel receptor γ predominantly expressed in skin. Nature 339:714–717

# *Growth Factors and the Control of Human Teratoma Cell Proliferation*

W. Engström, M. Tally, M. Granerus, E.P. Hedley, and P. Schofield

* Department of Pathology, Faculty of Veterinary Medicine, Swedish University of Agricultural Sciences, P.O. Box 7028, 75007 Uppsala, Sweden

## Introduction

Growth factors influence a wide range of mitotic and differentiative processes, which suggests that they might play a critical role in the control and integration of embryogenesis (cf. Engström and Heath 1988 for review). Development of the preimplantation mammalian conceptus can occur in the absence of exogenously supplied growth factors (Whitten 1970), which implied that either early embryonic stem cells lack the requirements for these factors or they fulfil them through endogenous synthesis in an auto- or paracrine fashion. One approach to mapping the range of potential growth factors and the spectrum of cellular responses one might expect from early human embryonic stem cells is to use pluripotential human teratoma cell lines as a model system (Andrews et al. 1984; Andrews 1985; Thompson et al. 1984; Engström et al. 1985; Engström 1986; Weima 1989). Such cells, like embryonic stem cells, have retained the ability to differentiate into a variety of tissues (Andrews 1984). Whilst such cell lines are clearly not equivalent to the ‘ES’ cell lines of the mouse, it has been recently argued (Pierce and Speers 1987; Rizzino et al. 1988) that growth factor expression by teratoma cells does not necessarily represent ectopic expression but rather reflects the normal production of growth factors by the tumour progenitor cells. This stance has gained profound support from comparison of the growth factors found in the early mouse conceptus and the analogous embryonal carcinoma cells of the mouse, which have been found to secrete a variety of multifunctional polypeptides, e.g. insulin-like growth factors (IGFs), platelet-derived growth factor (PDGF)-like activity, transforming growth factor (TGF)-like activity and heparin-binding growth factors such as basic fibroblast growth factor (bFGF). In the case of PDGF A chain and the transforming growth factors, at least, this pattern of synthesis has been shown to mirror blastocyst stage embryos (Rizzino 1985; Rizzino and Bowen-Pope 1985; Rizzino et al. 1988; Rappolee et al. 1988).

Recent Results in Cancer Research, Vol. 123

The insulin-like growth factors I and II are small peptide hormones with an approximate molecular weight of 7 kDa and are possibly the most abundant and ubiquitous in the human conceptus, eliciting a wide range of responses in almost all cell types studied, ranging from a mitogenic response to the induction or support of differentiation (Schofield 1988; Daughaday and Rotwein 1989). They are synthesised by a wide variety of tissue types during mammalian gestation (Scott et al. 1985; Hyldahl et al. 1986; Han et al. 1987; Brice et al. 1989; Ohlsson et al. 1989), and whilst there is clear evidence for the regulation of expression within these tissues the precise role of the IGFs in the differentiation or proliferation of particular tissues is unclear. In humans, levels of tissue expression of IGF II drop dramatically after birth (Enberg and Hall 1984), but are seen to be abnormally elevated in a number of childhood tumours, such as Wilms' tumours and hepatoblastoma (Scott et al. 1985; Haselbacher et al. 1987); this has been suggested to contribute to the neoplastic phenotype, as in the case of phaeochromocytoma (Haselbacher et al. 1987) and pituitary adenoma (Wilson et al. 1987). We have previously shown that primary human testicular tumours display elevated levels of IGF II transcript (Engström et al. 1986b). However, not all tumours share this characteristic and hence it has been of considerable interest to examine how other critical growth factor genes are expressed in this type of tumour. A typical survey of some primary embryonic tumours is shown in Fig. 1. In addition to the well-known heterogeneity in IGF II expression amongst germ cell tumours, it was found that whereas only the undifferentiated embryonal carcinoma contains basic FGF mRNA most germ cell tumours express the TGF β-1 gene.

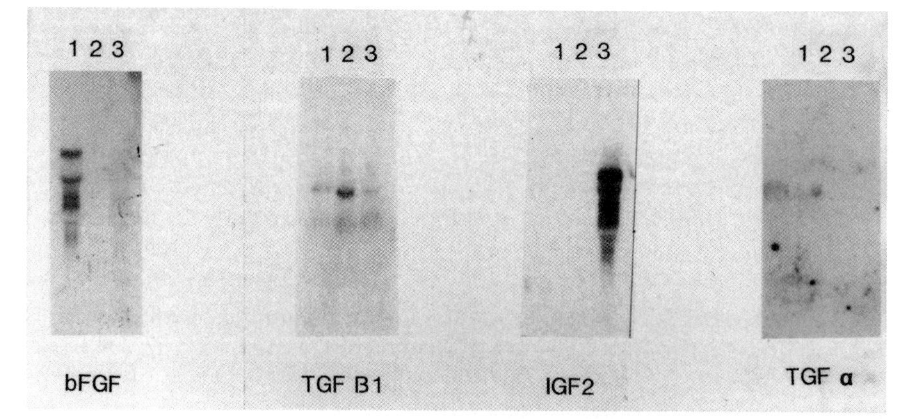


**Fig. 1.** Northern blot analysis of poly A+ RNA from three embryonic tumours (embryonal carcinoma, seminoma and Wilms' tumour). The analysis was performed exactly as described by Hyldahl et al. (1986). The filters were probed with radio-labelled cDNA probes for basic FGF, TGF-β 1, IGF II and TGF-α. 1, embryonal carcinoma; *2*, seminoma; *3*, WBSI Wiem's tumour

## Growth Cooperativity in the Early Embryo: Teratocarcinoma Cells as a Model System

Early evidence suggested the presence of a mutual growth-promoting effect of embryonic and extraembryonic components of the mouse conceptus. By using a refined organ-culturing technique, Gardner (1972) showed that the inner cell mass released some diffusible factor(s) which supports survival and growth of the trophectoderm, and vice versa. Similar growth cooperativity has been found in the growth of some teratocarcinoma cell types (reviewed by Heath and Rees 1985). In the mouse, teratocarcinomas are composed of two components: a malignant stem cell (embryonal carcinoma) and its differentiated progeny. It has proved possible to isolate a progressively growing component from such tumours which shows more or less differentiation (Rizzino 1983; Damjanov et al. 1987). In the case of the PC13 cell line, the progressively growing stem cells share characteristics with primitive ectoderm. However, it is clear that they are unable to give rise, in vivo, to all the components of a normal conceptus. In vitro, they can be induced to differentiate to give a population of primitive endoderm-like cells with progressively lengthening intermitotic periods.

Coculture of mouse embryonal carcinoma (EC) cells with their differentiated progeny (or with fibroblast target cells) not only leads to an enhanced survival of EC cells but also to the induction of heterologous target cell DNA synthesis and cell multiplication (Isacke and Deller 1983). It was initially suggested that one source of endodermlike-differentiated (END) cell growth-promoting substances was the undifferentiated parental EC cells. The factor responsible for stimulating END cell proliferation by their EC cell progenitors was termed embryonal carcinoma-derived growth factor (ECDGF) (Heath and Isacke 1984). ECDGF is one of many factors now isolated which display a strong affinity to heparin and hence has been grouped into the basic fibroblast growth factor family. Members of this family have been found in a variety of situations including the mammalian and amphibian embryos, where they have been implicated in the control of morphogenesis (Rosa et al. 1988; Godsave et al. 1988).

The reciprocal part of this relationship seems to comprise the synthesis of IGF II by the END cell population, which will stimulate the growth of the EC cells (Heath and Rees 1985). These findings suggest that a potential growth reciprocity might exist between the stem cell population and their differentiated progeny in the mouse embryo. This concept is valuable in that it would predict that the relative sizes of the stem and differentiated cell compartments would be dictated by the rate of transition between them, and hence by the rate of differentiation. However, evidence that it does indeed happen in the embryo is formally difficult to obtain.

**Proliferation of Human Teratoma Cells**

The human teratocarcinoma cell line Tera 2 was established from a pulmonary metastasis of a primary testicular teratocarcinoma (Fogh and Trempe 1975). Clones from this cell line can differentiate into at least two distinct cell types after exposure to retinoic acid or aggregation. These are characteristically marked by acquisition of HLA antigens and the expression of several other immunological markers (Thompson et al. 1984). One of the most easily identifiable cell types is neurons. These are a mixture of cholinergic and adrenergic types and express both neurofilament markers and tetanus toxin receptors. They bind the fibronectin made by the differentiated cells particularly strongly.

In vivo as a xenograft, Tera 2 produces many more cell types including obvious glandular epithelia (Thompson et al. 1984), and it is clear that, while Tera 2 shows restricted differentation, it is not pluripotent as would be expected by analogy to the mouse system. However, differentiated Tera 2 cells are not a progressively growing population and rapidly become refractory to serum or growth factor treatment after differentiation is induced. This would seem to be due to a fundamental switch set in train on differentiation as levels of c-*myc* remain high and characteristic of dividing cells in this quiescent population (Schofield et al. 1987). This was also observed for the differentiated components of primary teratoma (Sikora et al. 1985).

The stem cells of Tera 2 proliferate as small monomorphic cells in fetal calf serum. However, we have previously established that Tera 2 will proliferate in serum-free medium in the presence of IGF I or IGF II alone (Engström et al. 1985; Biddle et al. 1988). Kinetic S-phase labelling experiments demonstrate that IGF II acts on the survival of cells and the probability that they will complete another cell cycle having divided once, rather that on initiation of S phase. Furthermore, we were able to show that there exists a growth cooperativity between Tera 2 and its differentiated derivative analogous to mouse PC13 (Engström et al. 1986a). Because Tera 2 has such modes of growth factor requirements, it was interesting to ask whether it managed to fulfil its own growth factor requirements in an autocrine fashion.

We recently demonstrated that Tera 2 produces insulin-like growth factor 2 and releases it into the culture medium (Schofield et al. 1990). This is a property which the Tera 2 cells share with many other cell types of fetal and neoplastic origin (human fibrosarcoma cells (de Larco and Todaro 1978) and normal non-tumourigenic cells such as Buffalo rat liver cells (Dulak and Temin 1973), murine differentiated PC13 cells (Heath and Shi 1986), rat and human embryonic fibroblasts and Swiss mouse 3T6 cells (Engström and Tally, manuscript in preparation). Among other growth factors produced and released by tumour cells are the transforming growth factos (Todaro et al. 1976; Ozanne et al. 1980; Marquardt et al. 1983) and members of the sis-PDGF family (Betsholtz et al. 1983).

Our previous findings that Tera 2 cells will multiply in serum-free medium with IGF I or IGF II as the sole macromolecular supplement (Biddle et al. 1988) and that Tera 2 cells express large numbers of IGF II receptors (Engström et al. 1985; Weima 1989) suggest that we might expect an autocrine loop to be active in Tera 2 cell growth. However, this does not occur, and we propose that the reason for this is twofold. Firstly that Tera 2 releases two forms of IGF II (Schofield et al. 1990) and the major species of IGF II does not seem to be as active in the promotion of population growth as canonical IGF II, and secondly that Tera 2 produces a binding protein, which can inhibit the effects of IGFs on the cells. Not all the IGF II made by Tera 2 would appear to be complexed to this binding protein, as judged by the apparent molecular weight of IGF II on neutral gel filtration, which shows immunological and biochemical similarity to a previously characterised binding protein made by a wide variety of cell types in the human embryo (Hill et al. 1989). We suggest that these two properties contribute to reducing the mitogenic stimulus below a threshold necessary to promote growth. Similar proteins are released from a wide variety of tissues and animal cells in culture (Baxter and Martin 1989) and have been demonstrated to inhibit the effects of IGF I and IGFs on target cells (Drop et al. 1979).

The observation of cells which make growth factors, and have active receptors for them, yet fail to complete an autocrine loop, is not new and has been reported on several occasions for basic FGF. In this case it has been suggested that as bFGF lacks a consensus signal peptide it is not readily externalised and therefore cannot act on its receptors (reviewed by Schofield 1988). A similar situation exists in human breast carcinoma cells, which make both IGF I and its binding protein, yet still fail to grow in serum-free medium. However, autocrine or paracrine IGFs have been demonstrated to drive the proliferation of fetal liver cells in culture (Schofield 1988), and have been suggested to drive the proliferation of pituitary tumour cells in a xenograft context (Wilson et al. 1987). Consequently we must suppose that the mechanisms we see operating to prevent autocrine growth in Tera 2 are not always operative in cells producing the factor.

We recently demonstrated the appearance of two different IGF II species in Tera 2 cell-conditioned medium (Schofield et al. 1990). In addition to the 7-kDa canonical IGF II molecule, a "big" 14- to 15 kDa IGF II was identified by a combination of redioimmunoassay and radioreceptor assay. Whereas the two forms of IGF II were able to bind to placental membrane receptors and show immunological cross-reactivity, we found that "big" IGF II was significantly less mitogenic than canonical IGF II, assuming that the amounts of each protein measured using the same antibody were indeed comparable.

The presence of larger IGF II species has been described by others. Two variants were identified by Humbel and coworkers as a component of CSF (Zumstein et al. 1985); in one of them, Ser 29 of canonical IGF II has been replaced by the amino acids Arg-Leu-Pro-Gly. The other variant differs from 7-kDa IGF II by a substitution of Cys Gly Arg for Ser 33 and a carboxy-terminal extension of 21 amino acid residues (Zumstein et al. 1985). To exclude that "big" IGF II was identical to any of these forms, we analysed transcripts from the IGF II gene. Northern blotting revealed that Tera 2 cells produced the expected 6.0- and 4.8-k IGF II transcripts (Schofield et al. 1990). Furthermore, S1 nuclease analysis (Schofield and Tate 1987) showed that the majority of IGF II transcripts in Tera 2 are colinear over the alternatively spliced and variant regions of the IGF II gene. More recently, (Gowan et al. 1987) have demonstrated a high molecular weight form of IGF II from human plasma, which binds to IGF-binding protein and recognises IGF receptors on the placenta (Gowan et al. 1987). However, the two forms differ in biological properties, that found by Gowan et al. (1987) being more active than canonical IGF II on fetal fibroblasts, suggesting that the two variants may be structurally different.

Several mechanisms for generating a "big" IGF II can be envisaged. One possibility includes the appearance of an alternative splicing mechanism at the carboxy terminus [as with IGF I (Rotwein et al. 1986)], which can be effectively excluded on the basis of existing data. Another possibility is that "big" IGF II arises from aberrant or incomplete posttranslational processing, as has been suggested for the wide variety of forms of multiplication stimulating activity (MSA) produced during posttranslational processing (Yang et al. 1985).

It has been argued that growth factor expression by the stem cells of pluripotential tumours may caricature the precursor cell type, but nevertheless be a true reflection of its phenotype. Teratocarcinoma cells are thought to be derived from early embryonic stem cells and one might therefore expect the cells of the early human conceptus to show similar production and response to IGF II. Whether such cells produce qualitatively similar IGF II to Tera 2 is unknown, but a period of continuous culture or selection within a tumour context may be responsible for the unusual form of IGF (II) which we observe.

*Acknowledgements.* The authors would like to thank Professor C.F. Graham for his continued and enthusiastic support. This work was funded by the Cancer Research Campaign of Great Britain, the Riksforeningen mot Cancer, the Medical Research Council of Sweden and Barncancerfonden of Sweden. Invaluable assistance was given to the collaboration by the provision of an ACE award by the CIBA foundation.

## *References*

Andrews PW (1985) Properties of cloned embryonal carcinoma cells and their differentiated derivatives. Adv Biosci 55:64–71

Andrews PW, Damjanov I, Simon D, Banting GS, Carlin C, Dracupoli NC, Fogh J (1984) Pluripotent embryonal carcinoma cells derived from the human teratocarcinoma cell line Tera 2. Lab Invest 80:147–162

Baxter RC, Martin JL (1989) Binding proteins for the insulin like growth factors: structure, regulation and function. Prog Growth Factor Res 1:49–60

Betsholtz C, Heldin CH, Nister M, Ek B, Wasteson A, Westermark B (1983) Synthesis of a PDGF like growth factor in human glioma and sarcoma cells suggests the expression of the cellular homologue of the transforming protein of simian sarcoma virus. Biochem Biophys Commun 117:176–182

Biddle C, Li CH, Schofield PN, Tate VE, Hopkins B, Engström W, Huskisson NS, Graham CF (1988) Insulin like growth factors and the multiplication of Tera 2 – a human teratoma derived cell line. J Cell Sci 90:475–485

Brice A, Cheetham JC, Bolton V, Hill H, Schofield PN (1989) Temporal changes in the pattern of expression of the insulin like growth factor II gene associated with tissue maturation in the human fetus. Development 106:543–555

Damjanov I, Damjanov A, Solter D (1987) Experimental teratocarcinomas. In: Robertson E (ed) Teratocarcinoma and embryonic stem cells – a practical approach. IRL, Oxford

Daughaday WH, Rotwein P (1989) Insulin like growth factors I and II, peptides, mRNA, and gene structures, serum and tissue concentrations. Endocrine Rev 10:68–91

De Larco J, Todaro GD (1978) Growth factors from murine sarcoma virus transformed cells. Proc Natl Acad Sci (USA) 75:4001–4005

Drop SLS, Valiquette G, Guyda HJ, Corval MT, Posner BI (1979) Partial purification and characterisation of a binding protein for insulin-like growth factor activity. Acta Endocrinol (Copenh) 90:505–518

Dulak N, Temin HM (1973) Multiplication stimulating activity for chick embryo fibroblasts from rat liver cell conditioned medium. J Cell Physiol 81:161–170

Enberg G, Hall K (1984) Immunoreactive IGF II in serum of healthy subjects and patients with growth hormone disturbances and uraemia. Acta Endocrinol (Copenh) 107:164–170

Enberg G, Carlquist M, Jornvall H, Hall K (1984) The characterization of somatomedin A, isolated by microcomputer controlled chromatography reveals an apparent identity to insulin like growth factor I. Eur J Biochem 143:117–124

Engström W (1986) Differential effects of epidermal growth factor (EGF) on cell locomotion and cell proliferation in a cloned human embryonal carcinoma derived cell line in vitro. J Cell Sci 86:47–55

Engström W, Heath J (1988) Growth factors in early embryonic development. Perinatal Pract 9:11–32

Engström W, Rees AR, Heath JK (1985) Proliferation of a human embryonal carcinoma cell line in serum free medium; interrelationship between growth factor requirement and membrane receptor expression. J Cell Sci 73:361–373

Engström W, Heath JK, Rees AR (1986a) Growth phenotype of cloned human teratocarcinoma cells in vitro. Adv Biosci 55:77–78

Engström W, Hopkins B, Schofield P (1986b) Expression of growth regulatory genes in human testicular tumours. Int J Androl 10:79–84

Fogh J, Trempe S (1975) New human tumour cell lines. In: Fogh J (ed) Human tumour cell lines in vitro. Plenum, New York, pp 115–159

Gardner R (1972) An investigation of inner cell mass and trophoblastic tissues following their isolation from the mouse blastocyst. J Embryol Exp Morphol 28:279–312
Godsave SF, Isaacs HV, Slack JMV (1988) Mesoderm inducing factors – a small class of molecules. Development 102:555–567
Gowan L, Haptons B, Hill DJ, Schlueter RJ, Perdue J (1987) Purification and characterization of a unique high molecular weight form of insulin like growth factor II. Endocrinology 121:449–458
Han VKM, d'Ercole AJ, Lund PK (1987) Cellular localisation of IGF mRNA in the human fetus. Science 236:193–197
Haselbacher GK, Irminger, JC, Zapf J, Zeigler WH, Humbel RE (1987) Expression of IGF II in human adrenal phaeochromocytomas and Wilms' tumours; expression at the RNA and protein levels. Proc Natl Acad Sci USA 84:1104–1108
Heath JK, Isacke CI (1984) PC13 embryonal carcinoma derived growth factor. EMBO J 3:2957–2962
Heath JK, Rees AR (1985) Growth factors in mammalian embryogenesis. CIBA Found Symp 16:1–32
Heath JK, Shi WK (1986) Developmentally regulated expression of insulin like growth factors by differentiated murine teratocarcinomas and extraembryonic mesoderm. J Embryol Exp Morphol 95:193–212
Hill DJ, Clemmons DR, Wilson S, Han VKM, Strain AJ, Milner RDG (1989) Immunological distribution of one form of IGF binding protein and IGF peptides in human fetal tissues. J Mol Endocrinol 2:31–38
Hyldahl L, Engström W, Schofield PN (1986) Stimulatory effects of insulin like growth factors on the human embryonic cornea. J Embryol Exp Morphol 98:71–83
Isacke C, Deller MJ (1983) Teratocarcinoma cells exhibit growth cooperativity in vitro. J Cell Physiol 117:407–414
Marquardt H, Hunkapillar M, Hood LE, Twardzdik DR, de Larco J, Stephenson JE, Todaro GD (1983) Transforming growth factors produced by retrovirus transformed rodent fibroblasts and human melanoma cells. Proc Natl Acad Sci (USA) 80:4684–4688
Ohlsson R, Larsson E, Nilsson O, Wahlstrom T, Sundström P (1989) Blastocyst implantation precedes induction of insulin like growth factor II gene expression. Development 106:555–559
Ozanne B, Fulton BJ, Kaplan PL (1980) Kirsten murine sarcoma virus transformed cell lines and a spontaneously transformed rat cell line produce transforming growth factors. J Cell Physiol 105:163–180
Pierce GB, Speers WC (1987) Tumours as caricatures of the process of tissue renewal. Cancer Res 48:1996–2004
Rappolee DA, Brenner CA, Schultz R, Mark D, Werb Z (1988) Developmental expression of PDGF, TGFa and TGFB genes in preimplantation mouse embryos. Science 241:1823–1825
Rizzino A (1983) Two multipotential embryonal carcinoma cell lines irreversibly differentiate in defined media. Dev Biol 95:126–131
Rizzino A (1985) Early mouse embryos produce and release factors with transforming growth factor activity. In Vitro Cell Dev Biol 21:531–536
Rizzino A, Bowen Pope DF (1985) Production of PDGF like growth factors by embryonal carcinoma cells and binding of PDGF to their endoderm like differentiated cells. Dev Biol 110:15–22
Rizzino A, Kuszynski C, Ruff E, Tiesman J (1988) Production and utilisation of growth factors related to FGF by embryonal carcinoma cells and their differentiated derivatives. Dev Biol 129:61–71
Rosa F, Roberts AB, Danielpour D, Dart LL, Sporn MB, David IB (1988) Mesoderm induction in amphibians; the role of TGF beta 2 like factors. Science 239:783–784

Rotwein P, Pollack KM, Didier DK, Krivi GG (1986) Organisation and sequence of the human IGF I gene. Alternative processing produces two IGF I precursor peptides. J Biol Chem 261:4828–4832

Schofield PN, (1988) Growth factors in human embryogenesis. In: Jones CT (ed) Fetal and neonatal development. Perinatology Oxford, pp 24–31

Schofield PN, Tate VE (1987) Regulation of human IGF II transcription in fetal and adult tissue. Development 101:793–803

Schofield PN, Engström W, Lee AJ, Biddle C, Graham CF (1987) Expression of c-*myc* during differentiation of the human teratocarcinoma cell line Tera 2. J Cell Sci 88:57–64

Schofield PN, Tally M, Engström W (1990) Production of an enlarged form of IGF II by a human teratoma cell line. (submitted for publication)

Scott J, Cowell J, Robertson ME, Priestly JB, Wadey R, Hopkins B, Bell CI, Rall L, Graham CF, Knott TJ (1985) IGFII gene expression in Wilms' tumor and embryonic tissue. Nature 317:260–262

Sikora K, Evan G, Stewart J, Watson J (1985) Detection of the c-*myc* oncogene product in testicular cancer. Br J Cancer 52:171–176

Thompson S, Stern PL, Webb M, Walsh FS, Engström W, Evans EP, Shi WK, Hopkins B, Graham CF (1984) Cloned human teratoma cells differentiate into neuron like cells and other cell types in vitro. J Cell Sci 72:37–64

Todaro GF, de Larco JE, Cohen S (1976) Transformation by murine and feline sarcoma viruses specifically blocks binding of EGF to cells. Nature 264:26–31

Weima SM (1989) Growth factors and growth factor receptors in human testicular teratocarcinoma. PhD thesis, University of Utrecht

Whitton WK (1970) Nutrient requirements for the culture of preimplantation embryos in vitro. Adv Biosci 6:129–140

Wilson T, Thomas T, Haman T, Rosenfeld R (1987) Transplantation of IGF II secreting tumours into nude rodents. Endocrinology 120:1896–1901

Yang Y-H, Rechler MM, Nissley SP, Cologan JE (1985) Biosynthesis of rat IGF II. J Biol Chem 260:2578–2582

Zumstein PP, Luthy C, Humbel RE (1985) Amino acid sequence of a variant proform of insulin like growth factor II. Proc Natl Acad Sci (USA) 82:3169–3172

# *Role of Heparin-Binding Growth Factors in Embryonic Tumors*

E.J.J. van Zoelen

Department of Cell Biology, University of Nijmegen, Toernooiveld,
6525 ED Nijmegen, The Netherlands

## Introduction

Heparin-binding growth factors (HBGFs) form a rapidly expanding family of polypeptide growth factors, which are all characterized by a high affinity for heparin, and generally require heparin for activity. They have been shown to play a role in such diverse processes as cell proliferation, cell differentiation, cell survival, tumorigenesis, and angiogenesis (for recent reviews see Burgess and Maciag 1989; Baird and Walicke 1989; Rifkin and Moscatelli 1989). They exert their action by binding to high-affinity plasma membrane receptors, present particularly on mesodermal and epithelial cells, including neuronal and endothelial cell lines. One type of HBGF receptor has recently been cloned, and its sequence determined (Lee et al. 1989).

The best-characterized members of the HBGF family are acidic (a) and basic (b) fibroblast growth factor (FGF). In contrast to most other growth factors, the genes for both aFGF and bFGF do not contain a classical signal sequence, and therefore it is believed that FGFs are not secreted by cells, but remain localized within the cell (Thomas 1988). Recently, additional members of the HBGF family have been identified, in particular KFGF (Delli Bovi et al. 1987; Yoshida et al. 1987), INT-2 (Dickson and Peters 1987), FGF-5 (Zhan et al. 1988), FGF-6 (Marics et al. 1989), and keratinocyte growth factor (Rubin et al. 1989), often as products of potential cellular oncogenes. The genes for these additional HBGFs contain a signal sequence, which allows secretion of the above factors by the producer cells. Importantly, it has been shown that the presence of such a signal sequence largely contributes to the oncogenic properties of HBGFs (Rogelj et al. 1988; Blam et al. 1988).

Secreted HBGFs were first identified in medium conditioned by murine embryonal carcinoma (EC) cells, and many of the recently identified HBGFs are expressed during early stages of mammalian development. In this paper I will review the role of HBGFs in early embryonic development, and

Recent Results in Cancer Research, Vol. 123

correlate these observations to their role in murine and human terato-carcinomas.

Embryonal carcinoma cells are the undifferentiated stem cells of terato-carcinomas. In the mouse these tumors can be induced experimentally by explanting early embryos to an extrauterine site (Martin 1980). In humans these tumors occur spontaneously, particularly in the gonads (Andrews 1988). Because of the many properties they have in common with the pluripotent cells of the embryo itself, EC cells have been widely used as a model system for studying events involved in early embryonic differentiation and development (Martin 1980; Andrews 1988). More recently it has become possible to isolate inner cell mass cells of murine blastocysts directly in culture in an undifferentiated state, without an intervening tumor phase (Evans and Kaufman 1981; Martin 1981). These so-called embryonic stem (ES) cells have many properties in common with EC cells, and are believed to represent their normal embryonic counterparts very closely.

Recent studies on the development of *Xenopus laevis* have demonstrated that polypeptide growth factors play a direct role in cellular differentiation. Members of both the transforming growth factor (TGF)-β family (Kimelman and Kirschner 1987; Rosa et al. 1988) and of the HBGF family are capable of inducing mesoderm formation in ectodermal explants of competent embryos. With respect to the HBGF family such activity has been demonstrated for bFGF (Slack et al. 1987), aFGF (Grunz et al. 1988), KFGF, and INT-2 (Paterno et al. 1989), and for an embryonal carcinoma-derived growth factor (ECDGF) (Slack et al. 1987). A variety of polypeptide growth factors has been detected directly in early mouse embryos, including type α and type β TGFs, insulin-like growth factors (IGFs), and platelet-derived growth factor (PDGF), as reviewed elsewhere (Jakobovits 1986; Mercola and Stiles 1988; Heath and Smith 1989). TGF-α and members of the TGF-β and PDGF families have also been detected directly in early mouse blastula cells, but no evidence was found for expression of bFGF in these early developmental stages (Rappolee et al. 1988). In contrast, the *int-2* oncogene is known to be expressed during early murine development (Jakobovits et al. 1986), while bFGF has been detected in chicken embryos (Joseph-Silverstein et al. 1989). Unlike their direct differentiation-inducing effect during amphibian development, the role of polypeptide growth factors in murine development remains unclear.

## Characterization of Heparin-Binding Growth Factors Secreted by Murine Embryonal Carcinoma Cells

Undifferentiated murine EC cell lines, such as P19, PC13, and F9, are known to secrete a variety of polypeptide growth factors, including TGF-α, TGF-β, PDGFs, and IGFs (Jakobovits et al. 1986; Mercola and Stiles 1988; Rizzino and Bowen-Pope 1985; van Zoelen et al. 1989; Nagarajan et al. 1985), which underlines their similarity to early embryonic cells. In addition, however, these cells are among the best producers of secreted heparin-binding growth factors. In 1984, Heath and Isacke (1984) isolated a 17.5-kDa growth factor from PC13 EC-conditioned, serum-free medium designated ECDGF, which was unrelated to any other growth factor described at that time. In the same year, Shing et al. (1984) reported the purification of an endothelial cell growth factor from rat chondrosarcomas by the use of heparin-affinity chromatography. Following that observation it turned out that various polypeptide growth factors which had been identified in the early 1980s from a variety of tissues showed similar affinity to heparin, and were structurally closely related to either aFGF or bFGF (Burgess and Maciag 1989). In 1987

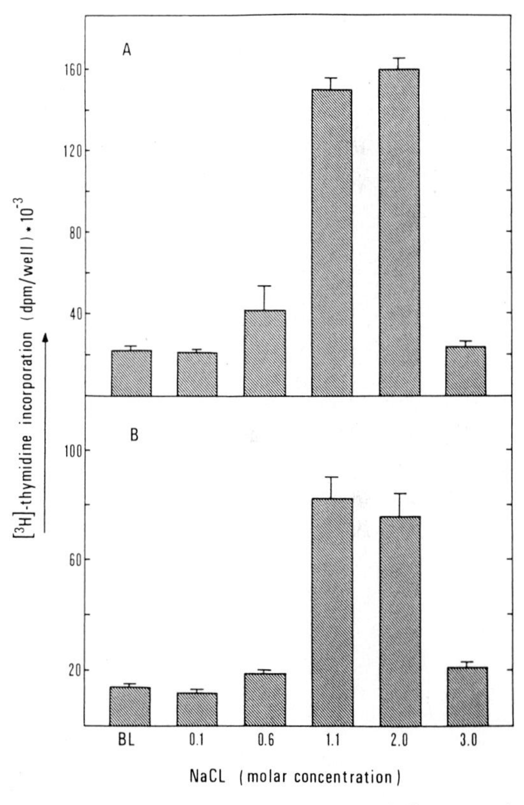


**Fig. 1A,B.** Heparin affinity chromatography of (**A**) PC13-derived growth factor and (**B**) basic FGF. Column was eluted stepwise by varying the NaCl concentration, and HBGF activity in the eluent was assayed for mitogenic activity ([$^3$H]thymidine incorporation) on quiescent Swiss 3T3 cells, as described by van Veggel et al. (1987). PC13-derived growth factor was first partly purified from PC13 EC-conditioned serum-free medium by elution from CM-Sephadex C-50 and phenyl-Sepharose C1-4B, as described by Heath and Isacke (1984). Basic FGF was otained from bovine brain (Collaborative Research). (From van Veggel et al. 1987)

we and others showed that PC13 EC-conditioned, serum-free medium contained a HBGF, with mitogenic activity for mesodermal cells (van Veggel et al. 1987; Heath 1987). Figure 1A shows that this PC13-derived growth factor elutes from heparin-Sepharose at a similar NaCl concentration to bFGF (Fig. 1B). On a CM-2-SW cation-exchange high-performance liquid chromatography (HPLC) column, it elutes at a slightly higher salt concentration than bFGF (Fig. 2), demonstrating that this PC13-derived growth factor is a more basic protein than bFGF (van den Eijnden-van Raaij et al. 1987). Similar differences in elution pattern have been described between ECDGF and bFGF (Heath and Smith 1989), indicating that this PC13-derived HBGF is most likely identical to ECDGF.

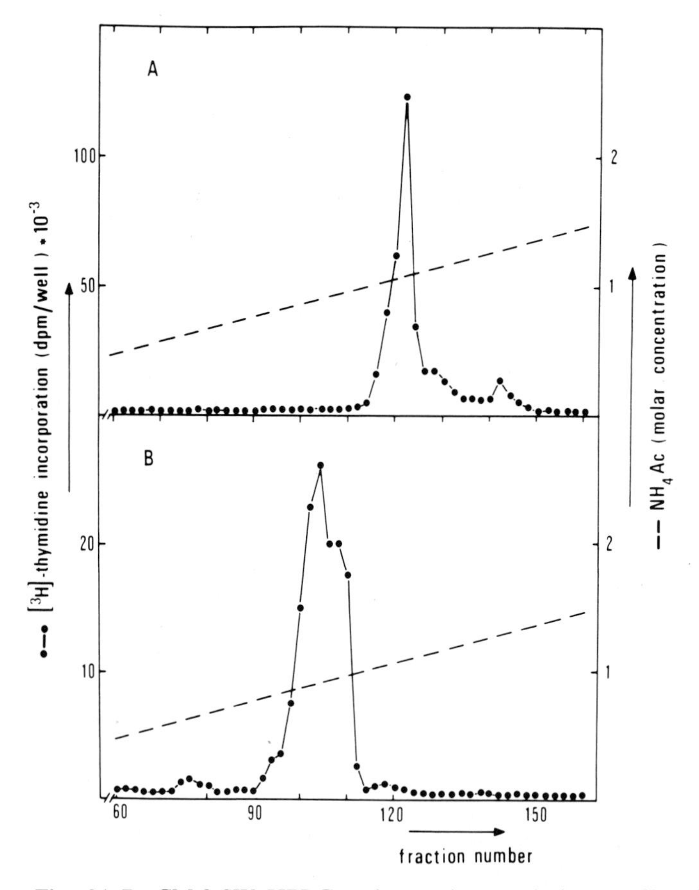


**Fig. 2A,B.** CM-2-SW HPLC cation-exchange elution profile of (**A**) PC13-derived growth factor and (**B**) basic FGF. Partly purified PC13-derived growth factor or pure bFGF was applied to the column in 1 *M* acetic acid, and eluted with a linear gradient (*dashed lines*) of $NH_4AC$ in 1 *M* acetic acid. Individual fractions were assayed for mitogenic activity ([$^3$H]thymidine incorporation) on quiescent Swiss 3T3 cells. (From van Veggel et al. 1987)

In order to assess that PC13 EC cells indeed secrete a HBGF, and that this factor is not released from dying cells under the above-used serum-free conditions, we have established a direct biological assay for HBGFs, using fetal bovine heart endothelial (FBHE) cells as a target cell line. FBHE cells essentially require HBGFs for proliferation, even when cultured in the presence of serum. This provides the possibility of testing for the presence of HBGFs in serum-containing medium, conditioned by EC cells during exponential proliferation. Figure 3 shows that PC13 EC-conditioned medium contains growth-stimulating activity for FBHE cells, only when tested in the additional presence of heparin. No such activity was observed in serum-containing medium conditioned by P19 EC cells (van Zoelen et al. 1989). Based on dose-response curves it was concluded that, on a quantitative basis, this HBGF is by far the most prominent polypeptide growth factor secreted by PC13 EC cells. Subsequent studies on the stability of HBGFs in solution demonstrated that in particular the oncogene-derived KFGF is rapidly inactivated in conditioned media, unless heparin is added during medium collection (Dell Bovi et al. 1988). When applied to P19 EC cells, this cell line appeared to secrete two different types of HBGF, one of which is immunologically related to KFGF, the other of which has properties similar to ECDGF (Heath et al. 1989). The exact relationship between ECDGF and the above-mentioned cloned genes of the HBGF family is still unclear.

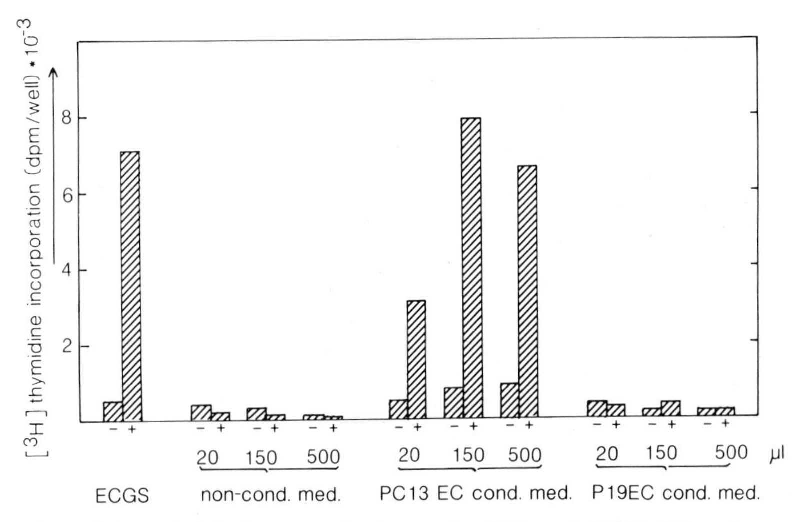


**Fig. 3.** Secretion of heparin-binding growth factors by P19 and PC13 EC cells. Growth-stimulating activity for fetal bovine heart endothelial cells ([$^3$H]thymidine incorporation) was determined in the absence (−) and presence (+) of 50 μg/m1 heparin, by adding the indicated volumes of serum-containing conditioned medium of these cell lines directly to the assay (total volume, 1.1 m1). Endothelial cell growth supplement (ECGS, 25 μg/m1) was used as a positive control in the assay. (From van Zoelen et al. 1989)

Presently, more and more information is appearing on the differential expression of *HBGF* genes during murine development. Braunhut et al. (1989) observed expression of the genes for both aFGF and bFGF in both undifferentiated and retinoic acid induced F9 cells, although a biologically active product could only be detected after differentiation. Rizzino et al. (1988) observed a decrease in production of FGF activity upon differentiation of both F9 and PC13 cells, accompanied by an increase in FGF receptor levels. Expression of the oncogene *int-2* is enhanced upon differentiation of both F9 and PCC4 cells (Smith et al. 1988; Yoshida et al. 1988a), but intriguingly expression of the gene for KFGF, which is located as a cluster with the *int-2* gene in the human genome (Wada et al. 1988), is decreased under these conditions. Also in ES cells derived directly from mouse embryos, expression of the *KFGF* gene is prominent (Heath et al. 1989), but in this case the effect of cellular differentiation on the expression level seems to depend on the nature of its differentiated progeny (Mummery, unpublished).

## Conclusion: Role of HBGFs in Human Testicular Teratocarcincomas

The observation that HBGFs are differentially expressed in differentiated and undifferentiated murine EC cells raises the question of whether HBGFs are involved in the pathology of human testicular tumors. Human EC cells, such as the Tera 2 cell line, differ from their murine counterparts in that they express a variety of polypeptide growth factor receptors, even in the undifferentiated state (Weima 1989). The observation that Tera 2 cells also secrete a variety of polypeptide growth factors (Weima et al. 1988) opens the possibility that proliferation of these cells is maintained by autocrine growth factor loops. With respect to HBGFs, NTera 2 cells have been shown to express the *bFGF* gene, and produce FGF activity (Tiesman et al. 1988). Expression of the *bFGF* gene has also been observed in undifferentiated Tera 2 cells (Mummery and Weima, unpublished), while differentiated Tera 2 cells express the *int-2* oncogene (Brookes et al. 1989). Recently we have observed that undifferentiated Tera 2 cells also express the gene for an FGF receptor (Claesson-Welsh and van Zoelen, unpublished). Since FGFs produced by cells remain mainly localized within the cytoplasm, coexpression of bFGF and an FGF receptor does not necessarily indicate the presence of an autocrine cycle, as indicated by data on a variety of nontransformed cells (Baird and Walicke 1989; Rifkin and Moscatelli 1989). Interestingly, however, recent data indicate that undifferentiated Tera 2 cells also express the gene for KFGF, while expression is lost after retinoic acid treatment (Mummery et al., unpublished). Because of the well-characterized oncogenic properties of KFGF (Delli Bovi et al. 1988), this observation indicates that secreted HBGFs may be involved in the tumorigenic growth behavior of human teratocarcinoma cells. This hypothesis is strengthened by the recent observation that the gene for KFGF is expressed in five out of nine surgically resected human testicular germ cell tumors, including seminomas and ECs (Yoshida et al. 1988b). It can be concluded that expression of the *KFGF* gene may well serve as an additional marker for the undifferentiated character of human testicular teratocarcinoma stem cells. We suggest that the secreted HBGFs play a functional role in the pathogenesis is these tumors.

## *References*


Andrews PW (1988) Human teratocarcinomas. Biochim Biophys Acta 948:17–36

Baird A Walicke PA (1989) Fibroblast growth factors. Br Med Bull 45:438–452

Blam SB, Mitchell R, Tischer E, Rubin JS, Silva M, Silver S, Fiddes JC, Abraham JA, Aaronson SA (1988) Addition of growth hormone secretion signal to basic fibroblast growth factor results in cell transformation and secretion of aberrant forms of the protein. Oncogene 3:129–136

Braunhut SJ, Gudas, LJ, Kurokawa, T, Sasse, J D'Amore, PA (1989) Expression of fibroblast growth factor by F9 teratocarcinoma cells as a function of differentiation. J Cell Biol. 108: 2467–2476

Brookes S, Smith R, Casey G, Dickson C, Peters G (1989) Sequence organization of the human *int-2* gene and its expression in teratocarcinoma cells. Oncogene 4, 429–436

Burgess WH, Maciag, T (1989) The heparin-binding (fibrobast) growth factor family of proteins. Annu Rev Biochem 58:575–606

Delli Bovi P, Curatola A, Kern F, Greco A, Ittman M, Basilico C (1987) An oncogene isolated by transfection of Kaposis's sarcoma DNA encodes a growth factor that is a member of the FGF family. Cell 50:729–737

Delli Bovi P, Curatola MA, Newmen KM, Sato Y, Moscatelli D, Hewick RM, Rifkin DB, Basilico C (1988) Processing, secretion, and biological properties of a novel growth factor of the fibroblast growth factor family with oncogenic potential. Mol Cell Biol 8:2933–2941

Dickson C, Peters G (1987) Potential oncogene product related to growth factors. Nature 326:833

Evans MJ, Kaufman MH (1981) Establishment in culture of pluripotential cells from mouse embryos. Nature 292:154–156

Grunz H, McKeehan WL, Knöchel W, Born J, Tiedemann H, Tiedemann H (1988) Induction of mesodermal tissues by acidic and basic heparin binding growth factors. Cell Differ 22:183–190

Heath JK (1987) Experimental analysis of a teratocarcinoma cell multiplication and purification of embryonal carcinoma derived growth factor. In: Robertson E (ed) Teratocarcinomas and embryonic stem cells: a practical approach. IRL, Oxford, pp 183–206

Heath JK, Isacke CM (1984) PC13 embryonal carcinoma derived growth factor. EMBO J 3: 2957–2962

Heath JK, Smith AG (1989) Growth factors in embryogenesis. Br Med Bull 45: 319–336

Heath JK, Paterno GD, Lindon AC, Edwards DR (1989) Expression of multiple heparin-binding growth factor species by murine embryonal carcinoma and embryonic stem cells. Development 107:113–122

Jakobovits A (1986) The expression of growth factors and growth factor receptors during mouse embryogenesis. In: Kahn P, Graf T (eds) Oncogenes and growth control. Springer, Berlin Heidelberg New York, pp 9–17

Jakobovits A, Shackleford GM, Varmus HE, Martin GR (1986) Two oncogenes implicates in mammary carcinogenesis, *int-1* and *int-2*, are independently regulated during mouse development. Proc Natl Acad Sci USA 83:7806–7810

Joseph-Silverstein J, Consigli SA, Lyser KM, Ver Pault C (1989) Basic fibroblast growth factor in the chick embryo: immunolocation to striated muscle cells and their precursors. J Cell Biol 108:2459–2466

Kimelman D, Kirschner M (1987) Synergistic induction of mesoderm by FGF and TGF-β and the identification of an mRNA coding for FGF in the early *Xenopus* embryo. Cell 51:869–877

Lee PL, Johnson DE, Cousens LS, Fried VA, Williams LT (1989) Purification and complementary DNA cloning of a receptor for basic fibroblast growth factor. Science 245:57–60

Marics I, Adelaide J, Raybaud F, Mattei MG, Coulier F, Planche J, de Lapeyrier O, Birnbaum D (1989) Characterization of HST-related FGF-6 gene, a new member of the fibroblast growth factor gene family. Oncogene 4:335–340

Martin GR (1980) Teratocarcinomas and mammalian embryogenesis. Science 209: 768–776

Martin GR (1981) Isolation of a pluripotent cell line from early mouse embryos cultured in medium conditioned by teratocarcinoma stem cells. Proc Natl Acad Sci USA 78:7634–7638

Mercola M, Stiles CD (1988) Growth factor superfamilies and mammalian embryogenesis. Development 102:451–460

Nagarajan L, Anderson WB, Nissley SP, Rechler MM, Jetten AM (1985) Production of insulin-like growth factor-II (MSA) by endoderm-like cells derived from embryonal cell growth. J Cell Physiol 124:199–206

Paterno GD, Gillespie LL, Dixon, MS, Slack JMW, Heath JK (1989) Mesoderm-inducing properties of INT-2 and KFGF: two oncogene-encoded growth factors related to FGF. Development 106:79–83

Rappolee DA, Brenner CA, Schultz R, Mark D, Werb Z (1988) Developmental expression of PDGF, TGFα and TGFβ genes in preimplantation embryos. Science 241:1823–1825

Rifkin DB, Moscatelli D (1989) Recent developments in the cell biology of basic fibroblast growth factor. J Cell Biol 109:1–6

Rizzino A, Bowen-Pope DF (1985) Production of PDGF-like growth factors by embryonal carcinoma cells and binding of PDGF to their endoderm-like differentiated cells. Dev Biol 110:15–22

Rizzino A, Kuszynski C, Ruff E, Tiesman J (1988) Production and utilization of growth factors related to fibroblast growth factor by embryonal carcinoma cells and their differentiated cells. Dev Biol. 129:61–71

Rogelj S, Weinberg RA, Fanning P, Klagsbrun M (1988) Basic fibroblast growth factor fused to a signal peptide transforms cells. Nature 331:173–175

Rosa F, Roberts AB, Danielpour D, Dart LL, Sporn MB, Dawid IB (1988) Mesoderm induction in amphibians: the role of TGF-β2-like factors. Science 239:783–785

Rubin JS, Osada H, Finch PW, Taylor WG, Rudikoff S, Aaronson SA (1989) Purification and characterization of a newly identified growth factor specific for epithelial cells. Proc Natl Acad Sci USA 86:802–806

Shing Y, Folkman J, Sullivan R, Butterfield C, Murray J, Klagsbrun M (1984) Heparin affinity: purification of a tumor-derived capillary endothelial cell growth factor. Science 223:1296–1299

Slack JMW, Darlington BG, Heath JK, Godsave SF (1987) Mesoderm induction in early *Xenopus* embryos by heparin-binding growth factors. Nature 326: 197–200

Smith R, Peters G, Dickson C (1988) Multiple RNAs expressed from the *int-2* gene in mouse embryonal carcinoma cell lines encode a protein with homology to fibroblast growth factors. EMBO J 7:1013–1022

Thomas KA (1988) Transforming potential of fibroblast growth factor genes. Trends Biochem. Sci 13:327–328

Tiesman J, Meyer A, Hines RN, Rizzino A (1988) Production of growth factors related to fibroblast growth factor and platelet-derived growth factor by human embryonal carcinoma cells. In Vitro Cell Devel Biol 24:1209–1216

Van den Eijnden-van Raaij AJM, Koornneef I, van Oostwarrd MJ, de Laat SW, van Zoelen EJJ (1987) Cation-exchange high-performance liquid chromatography: separation of highly basic proteins using volatile acidic solvents. Anal Biochem 163:263–269

Van Veggel JH, van Oostwaard TMJ, de Laat SW, van Zoelen EJJ (1987) PC13 embryonal carcinoma cells produce a heparin-binding growth factor. Exp Cell Res 169:280–286

Van Zoelen EJJ, Ward-van Oostwaard TMJ, Nieuwland R, van der Burg B, van den Eijnden-van Raaij AJM, Mummery CL, de Laat SW (1989) Identification and characterization of polypeptide growth factors secreted by murine embryonal carcinoma cells. Dev Biol 133:272–283

Wada A, Sakamoto H, Katoh O, Yoshida T, Yokota J, Little PFR, Sugimura T, Terada M (1988) Two homologous oncogenes, *hst-1* and *int-2*, are closely located in human genome. Biochem Biophys Res Commun 157:828–835

Weima SM (1989) Growth factors and growth factor receptors in human testicular teratocarcinoma. Thesis, University of Utrecht

Weima SM, van Rooijen MA, Mummery CL, Feijen A, Kruijer W, de Laat SW, van Zoelen, EJJ (1988) Differentially regulated production of platelet-derived growth factor and of transforming growth factor-beta by a human embryonal carcinoma cell line. Differentiation 38:203–210

Yoshida T, Miyagawa K, Odagiri H, Sakamota H, Little PFR, Terada M, Sugimura T (1987) Genomic sequence of *hst*, a transforming gene encoding a protein homologous to fibrobast growth factors and the *int-2*-encoded protein. Proc Natl Acad Sci USA 84:7305–7309

Yoshida T, Muramatsu H, Muramatsu T, Sakamoto H, Katoh O, Sugimura T, Terada M (1988a) Differential expression of two homologous and clustered oncogenes, *hst-1* and *int-2*, during differentiation of F9 cells. Biochem Biophys Res Commun 157:618–625

Yoshida T, Tsutsumi M, Sakamota H, Miyagawa K, Teshima S, Sugimura T, Terada M (1988b) Expression of the *hst-1* oncogene in human germ cell tumors. Biochem Biophys Res Commun 155:1324–1329

Zhan X, Bates B, Hu X, Goldfarb M (1988) The human FGF-5 oncogene encodes a novel protein related to fibroblast growth factors. Mol Cell Biol 8:3487–3495

# *Growth Factors and Receptors During Differentiation: A Comparison of Human and Murine Embryonal Carcinoma Cell Lines*

C.L. Mummery and S.M. Weima

Hubrecht Laboratory, Uppsalalaan 8, 3584 CT Utrecht, The Netherlands

## Introduction

Embryonal carcinoma (EC) cells are the pluripotential stem cells of teratocarcinomas, malignant tumors arising spontaneously in the gonads of humans and certain strains of mice, or induced experimentally by grafting early mouse embryos in extrauterine sites. Such tumors contain a disorganized mixture of undifferentiated stem cells together with apparently normal, mature cells and tissues arising from their differentiation. The stem cells are responsible for progressive growth of the tumor; if all differentiate, the tumors become benign and are termed "teratomas." Investigation of the mechanism of differentiation is thus intrinsic to understanding the reversibility of malignancy in these tumors. EC cells in culture have also been shown to be a particularly useful model system for studying the regulation of growth and differentiation during early murine development since they mimic both temporally and spatially many aspects of preimplantation and early postimplantational development (Graham 1977; Martin 1980). Their resemblance to early mouse embryos is most evident if they are cultured as aggregates in ascites fluid or *in vitro* under conditions where they are unable to attach to a substrate. The inner cells of the aggregate arrange as an epithelium resembling embryonic ectoderm while the outer cells differentiate to form an envelope of endoderm; the formation of these "embryoid bodies" thus resembles the earliest differentiation step in the inner cell mass (ICM) of a 5-day-old mouse embryo. The resemblance of EC cells to those of ICM is further illustrated by their ability to participate in normal development when introduced into a blastocyst subsequently reimplanted into a pseudopregnant female. Further, embryonal stem cells (ES cells) can be established in culture directly from the ICM of 5-day-old mouse embryos (Martin 1981; Evans and Kaufman 1981).

Both human and murine EC cells in culture possess many of the features of the transformed phenotype, including the ability to grow progressively and kill a suitable host, even at very low initial inocula (Rayner and Graham

Recent Results in Cancer Research, Vol. 123

1982). In addition to forming embryoid bodies, as described above, EC cells can be induced under a variety of other conditions *in vitro* to differentiate and form derivatives of all three germ layers. The cell types formed depend on the particular EC cell line, the nature and concentration of the inducer, and whether the inducer is added to cells cultured as aggregates or attached to a substrate. The most widely used inducer, effective in most EC cell lines, is retinoic acid (RA) (Strickland and Mahdavi 1978; Hogan et al. 1981; Jones-Villeneuve et al. 1982; Edwards et al. 1983). The differentiation of EC cells is marked by major changes in cell phenotype 24–48 h after induction; altered morphology, reduction in growth rate, and loss of tumorigenic potential are the most striking changes observed. The cells also lose expression of several EC-specific markers, such as stage-specific embryonic antigen 1 (SSEA-1) (Solter and Knowles 1978) and alkaline phosphatase (Bernstine et al. 1973) and begin to express certain proteins characteristic of differentiated cells such as laminin (Strickland and Mahdavi 1978; Adamson et al. 1979; Wang and Gudas 1983), plasminogen activator (Strickland and Mahdavi 1978), and epidermal growth factor (EGF) receptors (Rees et al. 1979).

Although murine and human EC cells show a morphological resemblance, they differ significantly in their differentiation potential. Trophectodermal elements are frequently found associated with human teratocarcinomas in vivo but are rarely found in murine tumors; however, given the relationship established between murine EC cells and cells of the ICM, which have generally lost the capacity to form trophectoderm, such elements would not be expected. Even though the EC cells of human teratocarcinomas form well-differentiated tumors containing endodermal, mesodermal, and ectodermal derivatives when injected into athymic (nu/nu) mice, they show only limited spontaneous differentiation in vitro (Andrews et al. 1980, 1983a, b, 1984). Relatively few of these cell lines respond to RA in vitro, but both endoderm-like and neuron-like cells can be induced under appropriate conditions (Andrews et al. 1984b; Thompson et al. 1984). Human EC cells do not express the SSEA-1 antigen but do often express SSEA-3 (Andrews et al. 1982), a carbohydrate carried by either membrane glycolipids or glycoproteins (Shevinsky et al. 1982) and expressed on mouse oocytes and cleavage stage embryos until early blastocyst. Human EC cells have been observed to become SSEA-3 negative and SSEA-1 positive after differentiation by low-density plating (Andrews et al. 1984a). These data have led Andrews et al. (1982, 1984) to suggest that the presumptive human stem cells represent a precleavage stage cell type, i.e., an earlier developmental stage than their murine counterparts. On the other hand, some human EC cell lines, such as the clonal lines derived from Tera 2, do express EGF receptors (Engström et al. 1985; Carlin and Andrews 1985), normally expressed by murine EC cells only after differentiation (Rees et al. 1979). Further, while the differentiated derivatives of murine EC cells are no longer able to form tumors (Rayner and Graham 1982), the extraembryonic and somatic elements of human teratocarcinomas may still be malignant.

In summary therefore murine EC cells provide an excellent system for studying the loss of the transformed phenotype of an embryonic tumor under controlled conditions in vitro and the mechanisms of action of anti-tumorigenic substances such as RA; in addition it provides a frame of reference for investigating similar processes in human material.

## Growth Properties of Human Versus Murine EC Cells During Differentiation

In the studies we will describe, the human cell line Tera 2 cl 13 (further referred to as Tera 2) has been compared with the murine EC cell lines PC13, P19, and F9. All of these cell lines differentiate in monolayer in response to RA to form flattened cell types, mostly with endoderm-like properties. The morphology of Tera 2 cells 5 days after addition of RA compared with the undifferentiated cells is shown in Fig. 1. We have shown previously by analysis of cell cycle kinetics that while in murine EC cells growth rate is only reduced two cell cycles after RA addition (Mummery et al. 1984), exponentially growing Tera 2 cells respond considerably more rapidly (Table 1) and the cells accumulate in the G1/G0 phase of the cell cycle (Mummery et al. 1987). It is not clear whether this is related to the intrinsically longer

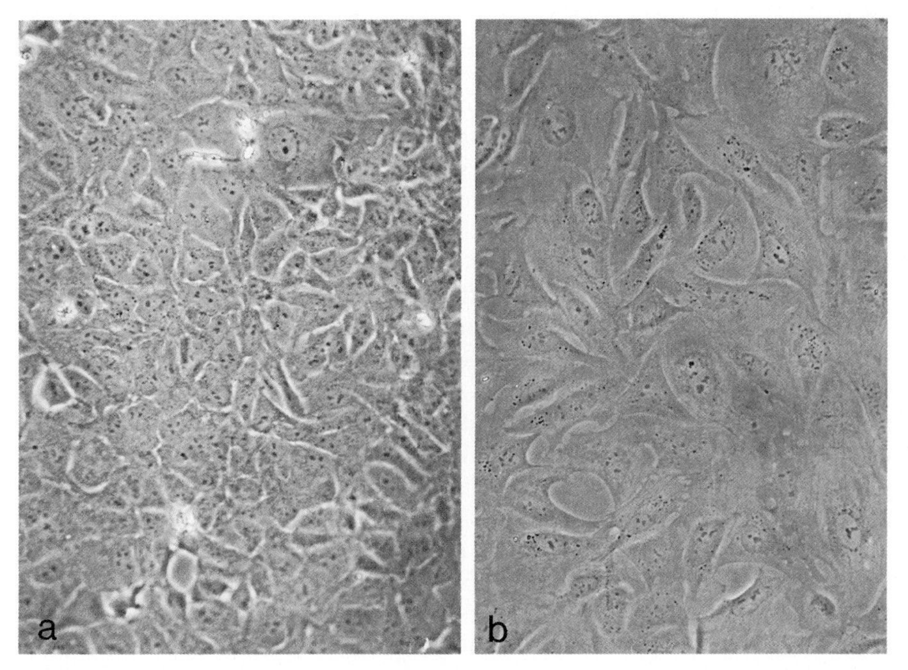


**Fig. 1a,b.** Effect of RA on the morphology of Tera 2 in monolayer (**a**) control cultures and (**b**) cultures 5 days after the addition of RA ($5 \times 10^{-6}$ M). (Mummery et al. 1987)

**Table 1.** Effect of RA on the intermitotic times of Tera-2 compared with PC13 and P19 EC cells

| Cell | Intermitotic times, $T_i$ (h) | | | |
|---|---|---|---|---|
| | Control (without RA) | 1st generation | 2nd generation | 3rd generation |
| Tera-2 | $23.3 \pm 0.3$ | $30.7 \pm 1.6$ | $31.9 \pm 1.4$ | $30.8 \pm 1.6$ |
| P19 | $12.1 \pm 0.4$ | $12.9 \pm 0.8$ | $11.0 \pm 0.7$ | $17.9 \pm 0.8$ |
| PC13 | $11.6 \pm 0.2$ | $12.3 \pm 0.7$ | $12.2 \pm 0.3$ | $15.8 \pm 0.3$ |

Data from Mummery et al. (1987).
RA was added at time zero and the mean intermitotic times for each of the three following generations determined directly from film data.

doubling time of the human EC cells but illustrates one aspect of their dissimilarity with murine EC cells.

## Growth Factor Expression During Differentiation

The growth of undifferentiated EC cells in culture in the absence of exogenous growth factors may be due to their ability to synthesize the factors they require themselves and remain in a continuous state of growth by an autocrine mechanism. Several studies have shown that EC cells and mouse embryos produce a variety of polypeptide grwoth factors including type α and type β transforming growth factors (TGFs), insulin-like growth factors (IGFs), platelet-derived growth factor (PDGF), and heparin-binding growth factors (HBGFs) (reviewed Mummery and van den Eijnden-van Raaij 1990; see also van Zoelen, this volume). We screened mRNA isolated from Tera 2 cells before and after differentiation for the expression of members of these growth factor families by Northern blot analysis and compared this with the expression in P19 EC cells during differentiation. Figure 2 shows the results obtained with probes for PDGF-A and PDGF-B (analogous to the oncogene c-*sis*), TGF-$\beta_1$, -$\beta_3$, and -$\beta_4$, and IGF II in P19 cells. Both PDGF-A and PDGF-B are induced within 24 h of RA addition while individual members of the TGF-β family are clearly independently regulated. With the cDNA probe used, TGF-$\beta_1$ is expressed as transcripts of 2.5 and 1.8 kb, the latter being greatly reduced by the induction of differentiation. We cannot exclude that this actually represents TGF-$\beta_4$ (Fig. 2) as the transcripts are the same size and homology between TGF-$\beta_1$ and TGF-$\beta_4$, makes the possibility of cross-hybridization high. We have shown previously that TGF-$\beta_2$ is not produced by P19 EC cells but both the RNA and protein are detectable after differentiation (Mummery et al. 1990) IGF I was not detectable either before or after differentiation (not shown) while IGF II is induced within 24 h of RA addition, reaching a maximum after 4–5 days when cells appear terminally

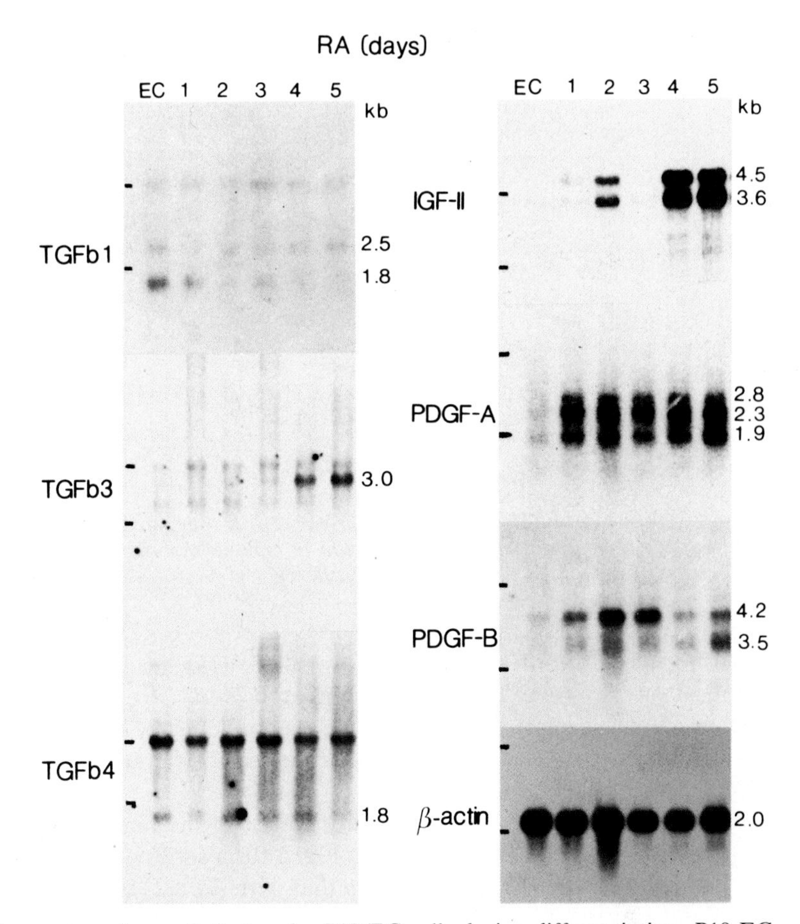


**Fig. 2.** Expression of growth factors by P19 EC cells during differentiation. P19 EC cells were induced to differentiate in monolayer by the addition of RA for the time indicated. RNA was isolated and poly-A$^+$ RNA was selected and analyzed by Northern blotting as described previously. (Mummery et al. 1990)

differentiated. Similar experiments in Tera 2 show that, in contrast to P19, PDGF-A is expressed in the undifferentiated cells and is reduced by RA while PDGF-B is undetectable either before or after differentiation (Fig. 3; from Weima et al. 1989b); in the expression of PDGF-A, Tera 2 cells show a greater resemblance to F9 EC than P19 EC (Mummery et al. 1989). As in all murine EC cells, the cDNA probe for TGF-$\beta_1$ recognizes two transcripts in Tera 2 EC but in this case the larger transcript is greatly increased by differentiation and the smaller transcript unchanged (Fig. 4a; from Weima et al. 1989b). This again could represent TGF-$\beta_4$ although this is barely detectable on Northern blots (Fig. 4b). As in all murine EC cells, TGF-$\beta_2$ expression is increased by differentiation (Fig. 4a; from Weima et al. 1989c)

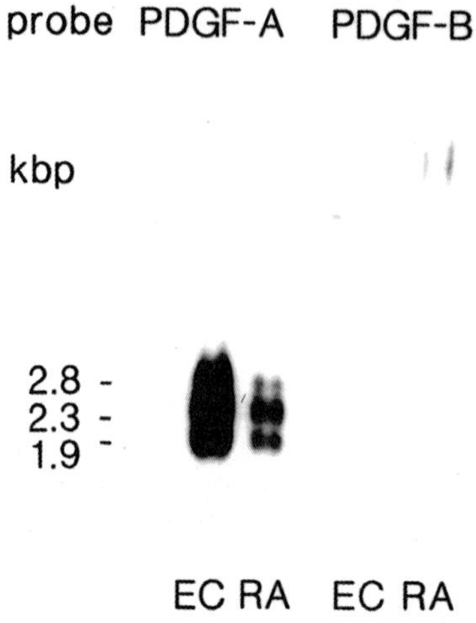

probe PDGF-A PDGF-B

kbp

2.8 -
2.3 -
1.9 -

EC RA EC RA

**Fig. 3.** Expression of PDGF-A and PDGF-B during the differentiation of Tera 2. Differentiation was induced by RA addition to cells in monolayer, as in Fig. 1, and RNA isolated as in Fig. 2. (Weima et al. 1989a)

while no hybridization to the TGF-$\beta_3$ probe was detectable (Fig. 4b). Biochemical studies revealed that differentiated Tera 2 cells produce TGF-β protein of which approximately 25% is in a biologically active form. Moreover, immunoreactive TGF-$\beta_2$ was detectable in differentiated Tera 2 cells (Weima et al. 1989c).

Finally, the expression of kFGF, a member of the family of HBGFs (see van Zoelen, this volume), is shown in Fig. 5 for a series of differentiated and undifferentiated EC and ES cells, together with Tera 2. In all cases, kFGF is expressed in the undifferentiated cells and is reduced or undetectable after differentiation.

**Table 2.** Growth factor expression compared in human and murine EC cells

| | Tera 2 | | P19 | | F9 | |
|---|---|---|---|---|---|---|
| | EC | +RA | EC | +RA | EC | +RA |
| PDGF A | ++ | + | – | ++ | ++ | + |
| PDGF B | – | – | +/– | ++ | nd | nd |
| TGF β1 2.5 Kb | + | ++ | + | + | + | + |
| 1.8 Kb | + | + | ++ | – | ++ | + |
| TGF β2 | – | ++ | – | ++ | – | + |
| TGF β3 | – | – | – | + | nd | nd |
| TGF β4 | +/– | +/– | + | +/– | nd | nd |
| IGF I | – | – | – | – | – | + |
| IGF II | – | – | – | ++ | – | ++ |
| kFGF | ++ | – | ++ | – | ++ | – |

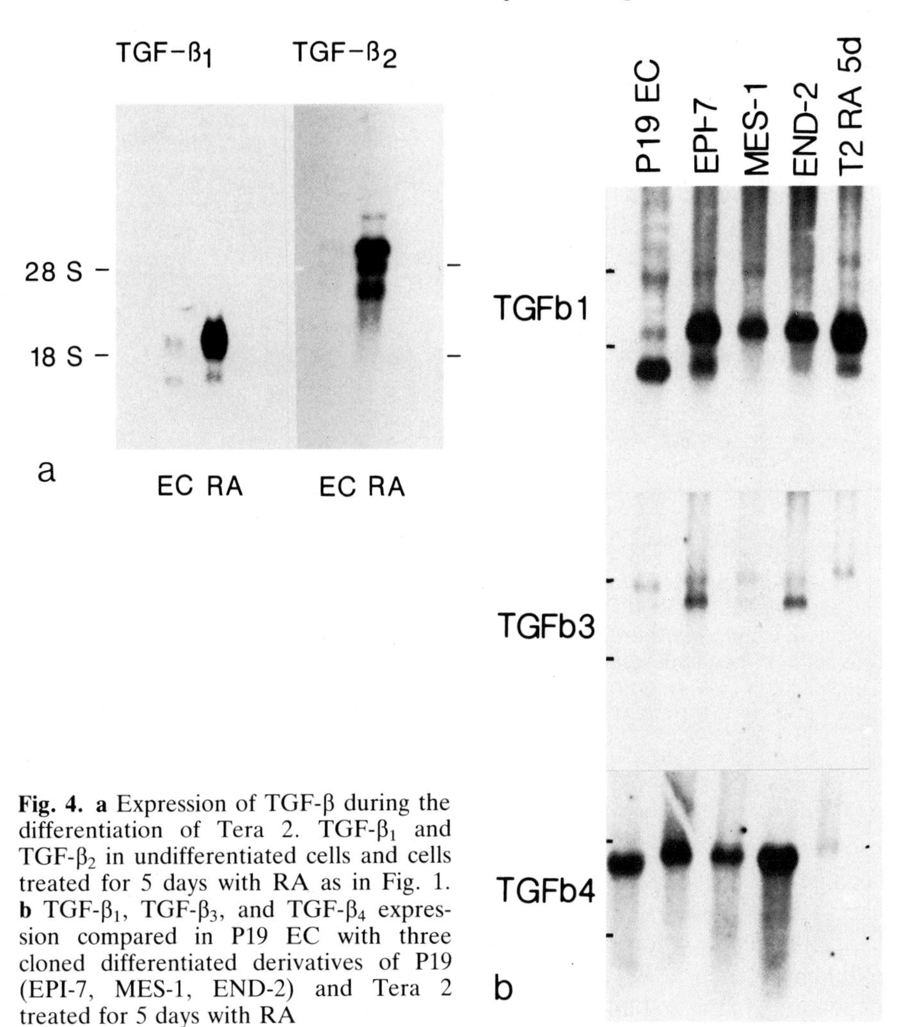


**Fig. 4.** **a** Expression of TGF-β during the differentiation of Tera 2. TGF-$\beta_1$ and TGF-$\beta_2$ in undifferentiated cells and cells treated for 5 days with RA as in Fig. 1. **b** TGF-$\beta_1$, TGF-$\beta_3$, and TGF-$\beta_4$ expression compared in P19 EC with three cloned differentiated derivatives of P19 (EPI-7, MES-1, END-2) and Tera 2 treated for 5 days with RA

These expression studies in Tera 2 cells are summarized and compared with P19 and F9 EC in Table 2. In general Tera 2 cells resemble (some) murine EC cell lines during differentiation in the expression of PDGF-A TGF-$\beta_1$, kFGF, and IGF I and differ in the expression of PDGF-B, TGF-$\beta_1$, -$\beta_4$, and IGF II. At present we have insufficient data to draw conclusions about TGF-$\beta_3$.

## Growth Factor Receptor Expression During Differentiation

Like many other tumor cell lines grown in vitro, human and murine EC cells are essentially independent of polypeptide growth factors (Weima et al. 1989a; Heath and Deller 1983; Engström et al. 1985). However, when they

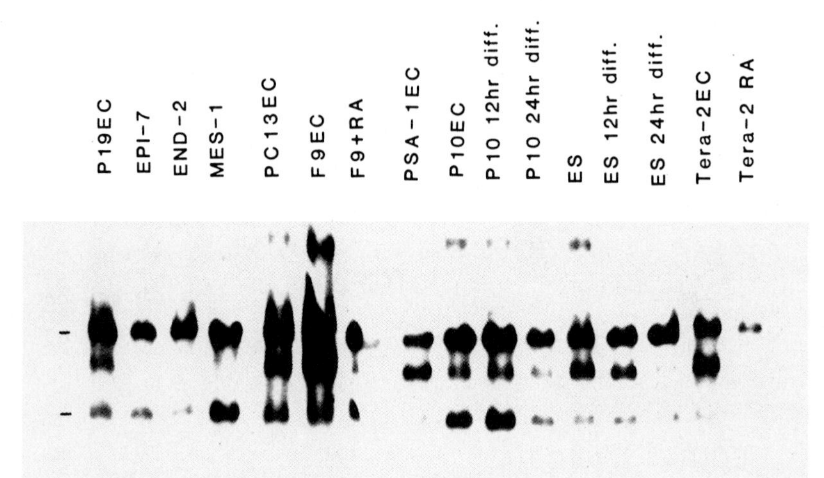


**Fig. 5.** Expression of kFGF during the differentiation of EC and ES cells. Expression is shown for a series of murine EC cell lines both feeder dependent (PSA-1, P10) and independent (P19, PC13, F9) and some of their differentiated derivatives, as indicated. Changes within 24 h of RA addition to murine ES cells are also shown. Tera 2 cells were either undifferentiated or fully differentiated by 5 days addition of RA. The specific kFGF transcript lies between the 28 s and 18 s bands indicated

are grow in serum-free madia, a dependency on IGF becomes manifest (Heath and Deller 1983; Biddle et al. 1988). In murine EC cells the effect on proliferation of a particular growth factor correlates with the expression of its cognate cell surface receptor, i.e., they do not possess growth factor receptors except for IGFs (Heath and Shi 1986). Upon differentiation of murine EC cells, receptors for EGF (Rees et al. 1979; Mummery et al. 1985), PDGF (Mummery et al. 1986) and/or TGF-β (Rizzino 1987) become expressed depending on the cell line and the induction protocol used. As a consequence the growth properties of the differentiated derivatives can be modulated by the growth factors recognizing these receptors. In contrast the human EC cell lines studied so far express EGF receptors when they are in an undifferentiated state, while they lose these receptors when they are induced to differentiate with RA (Carlin and Andrews 1985; Engström et al. 1985; Weima et al., unpublished). Despite the finding of functional EGF-dependent kinase activity in these cells, a significant effect of EGF on the proliferation of human EC cells has never been described. The functional IGF I receptors expressed by Tera 2 cells (Weima et al. 1989c) appear to have a role in mediating the effect of IGFs and insulin on the proliferation of Tera 2 cells (Biddle et al. 1988).

### *TGFβ Receptors*

High-affinity cell surface TGF-β-binding proteins, operationally defined as TGF-β receptors, have been implicated in TGF-β action and described by several investigators (Cheifetz et al. 1987; Ewton et al. 1988; Segarini and Seyedin 1988). Type I, II, and III receptors, 55-kDa, 75-kDa, and 300-kDa glycoproteins, respectively, have been proposed to be involved in TGF-β action. Studies with TGF-β-resistant mutants showed that the expression of type I receptors is essential for the effect of TGF-β on extracellular matrix deposition and cellular proliferation (Boyd and Massagué 1989). The growth of mouse EC cell lines F9 and PC13 is unaffected by TGF-$\beta_1$ or TGF-$\beta_2$ unless the cells are induced to differentiate (Rizzino 1987). It was found that these undifferentiated cells lack specific high-affinity binding sites for TGF-β, whereas TGF-β receptors are induced by treatment with RA for a period as short as 48 h. In addition, the growth of P19 EC cells is unaffected by TGF-β while EPI-7 cells, a differentiated P19-derived cell line, possess a TGF-β-sensitive phenotype (van Zoelen et al. 1989).

Analysis of the data presented in Fig. 6 shows that undifferentiated and differentiated Tera 2 cells express 14000 and 1800 TGF-β-binding sites respectively. Similar data were obtained using NTera 2 clone D1 cells. However, using the same assay as Rizzino (1987) to monitor the effect of TGF-β on the proliferation of EC cells and EC cells induced to differentiate, we found no effect of TGF-$\beta_1$ or of TGF-$\beta_2$ on the growth of Tera 2 cells (not shown). A structural analysis of TGF-β receptors by means of affinity labeling, as shown in Fig. 7 (from Weima et al. 1989b), revealed that undifferentiated Tera 2 cells express type III TGF-β receptors (200–300 kDa) but lack the type I (50-kDa) and II (75-kDa) receptors; these are readily detectable in the NRK cells used as a positive control. As outlined above, the lack of a TGF-β-responsive phenotype might be attributed to the absence of type I TGF-β receptors. The number of TGF-β-binding sites on differentiated Tera 2 cells is apparently below the detection limit of this assay. In conclusion, although the growth of both human and murine EC cells is unaffected by TGF-β, human EC cells do express TGF-β receptors while their murine counterparts do not. In addition, while RA induces the appearance of TGF-β receptors in murine systems, the number of TGF-β receptors in human cells is decreased to very low levels.

### *Platelet-Derived Growth Factor Receptors*

Two types of transmembrane PDGF receptors have been identified, a so-called type A receptor which binds all forms of PDGF and a type B receptor which binds B-chain-containing dimers only, i.e., PDGF-AB and PDGF-BB (C.H. Heldin et al. 1988; Hart et al. 1988). The primary structures of type A (Matsui et al. 1989) and type B PDGF receptor (Yarden et al. 1986; Gronwald et al.

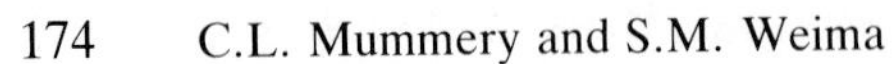


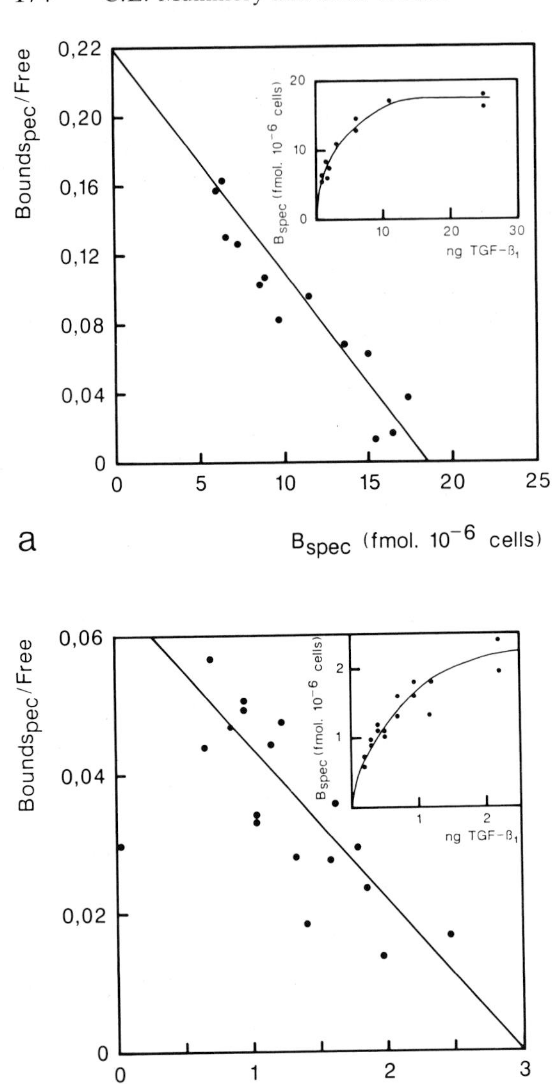


**Fig. 6 a,b.** Binding of TGF-$\beta_1$ to undifferentiated (**a**) and differentiated (**b**) Tera 2 cells (see Fig. 1). Subconfluent monolayers of cells were incubated with a fixed concentration of iodinated TGF-$\beta_1$ and increasing concentrations of unlabeled TGF-$\beta_1$ in hydroxyethylpiperazine ethanesulfonic (HEPES)-buffered acid Dulbecco's minimum essential medium (DMEM) containing 1 mg/ml BSA for 3 h at room temperature. After washing of the cells bound radioactivity was extracted with 1% Triton X-100 and determined in a gamma counter. The radioactivity bound in the presence of 100-fold excess unlabeled TGF-$\beta_1$ was subtracted from the data presented in the inset. Data are plotted according to Scatchard (1949)

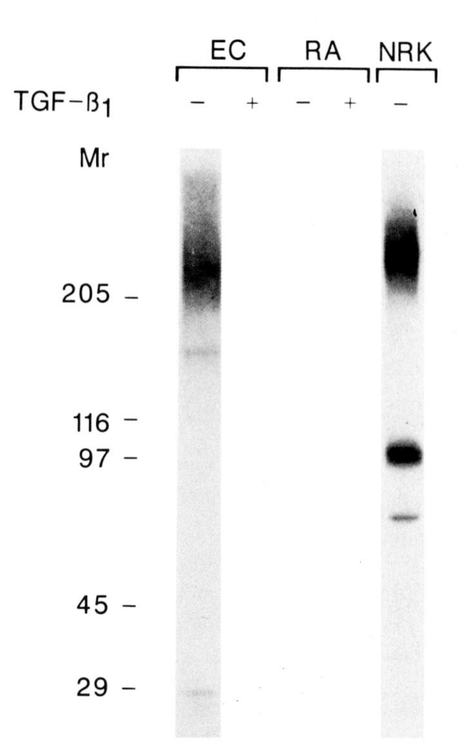


**Fig. 7.** Affinity labeling of TGF-β receptors. Monolayers of undifferentiated (*EC*) and differentiated (*RA*) Tera 2 cells and normal rat kidney fibroblasts. (*NRK*) were incubated with 2.5 ng/ml iodinated TGF-$\beta_1$ for 3.5 h at 4°C; where indicated (+) in the presence of 0.5 µg/ml unlabeled TGF-$\beta_1$. After washing the cells, bound TGF-β was covalently linked to the cells by incubating them with dioctyl sodium Sulfosuccinate (DSS). TGF-β receptors were subsequently solubilized and analyzed by electrophoresis on 5%–15% SDS-polyacrylamide gels and autoradiography. The molecular masses of the standards are given in kilodaltons

1988) have been revealed by cDNA cloning, and contain 1089 and 1098 amino acids respectively. Common structural features include the presence of a long sequence that interrupts the tyrosine-specific protein kinase of these molecules. In rodent cell systems the precursor of the type A receptor is consistently smaller than the type B receptor precursor, 160 kDa and 165 kDa respectively (Matsui et al. 1989). Precursor and mature type A PDGF receptors synthesized by human fibroblasts and glioma cells are 140 and 175 kDa respectively, whereas the corresponding type B proteins are 160 and 185 kDa, respectively (Claesson-Welsh et al. 1987, 1989). The molecular weights of fully processed cell surface exposed PDGF receptors are dependent on the cell type and are most likely determined by differences in glycosylation. Type B receptors are expressed predominantly in cells of mesodermal origin although aberrant expression in epithelial cells has been described (N.E. Heldin et al. 1988). Binding of PDGF to its receptor leads to

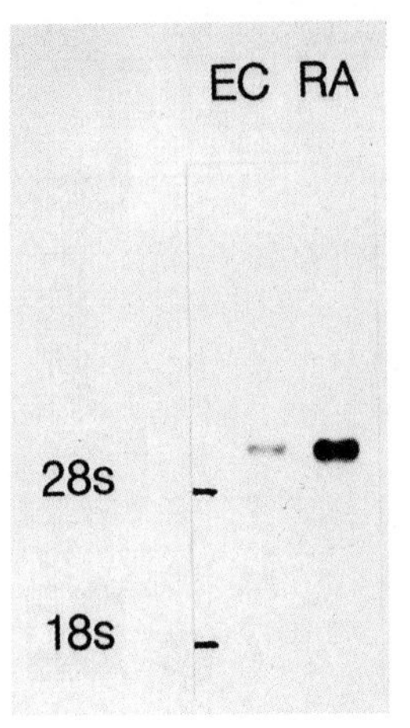


**Fig. 8.** Platelet-derived growth factor receptor mRNA expression by Tera 2 cells. Poly-($A^+$)-RNA isolated from undifferentiated cells (*EC*) and cells treated for 6 days with *RA* were hybridized with a human type-B PDGF receptor cDNA probe. Molecular size of the markers is expressed in Sedberg units.

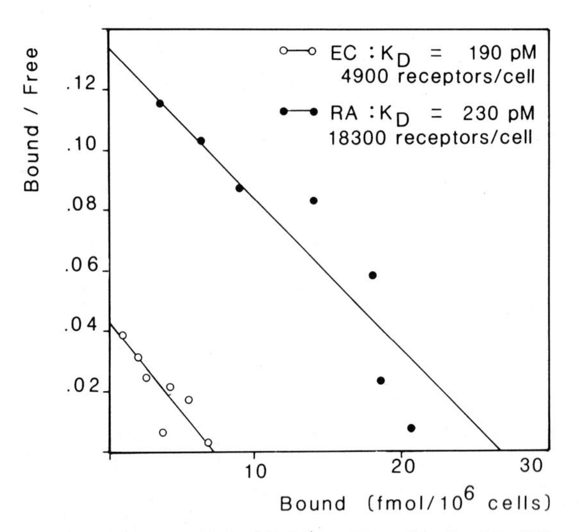


**Fig. 9.** Binding of PDGF-BB to Tera 2 cells. Undifferentiated (*EC*) and differentiated (*RA*) Tera 2 cells were grown to near confluence and incubated overnight in a growth factor-free medium. The cells were incubated for 1 h at room temperature with iodinated PDGF-BB and increasing concentrations of unlabeled PDGF-BB in binding medium. Bound radioactivity was subsequently extracted with 1% Triton X-100. Nonspecific binding was determined by the linear subtraction method (van Zoelen 1989) and data for specific binding were plotted according to Scatchard (1949)

a series of so-called early responses which have been implicated in the induction of DNA synthesis and cell division. These responses include activation of the receptor tyrosine kinase, autophosphorylation of the receptor on tyrosine residues, and rapid internalization and degradation of the ligand-receptor complexes (reviewed by Williams 1989). However, the biochemical pathways leading to DNA synthesis remain to be elucidated (Williams 1989).

For the murine EC cell lines F9 and PC13 it has been established that the expression of PDGF receptor is a typical feature of differentiated endoderm-like cells (Rizzino and Bowen-Pope 1985). In the more pluripotent P19 cell line, PDGF receptors become expressed along differentiation pathways leading to mesoderm (Mummery et al. 1986). The type of PDGF receptor in these cells remains to be established.

Figure 8 (from Weima et al. 1990) shows that undifferentiated Tera 2 cells express low but significant levels of type B PDGF receptor mRNA. The levels of mRNA are markedly increased after induction of differentiation with RA. The difference in mRNA levels parallels the number of receptors expressed by differentiated and undifferentiated Tera 2 cells (Fig. 9). However, the data presented in Fig. 9 can be the result of binding to both type and type B receptors, since PDGF-BB which binds to both types of receptors was used as a ligand. Figure 9 (from Weima et al. 1990) presents the results of a typical pulse-chase experiment showing the effect of different PDGF isoforms, i.e., AA and BB, on PDGF receptor turnover in Tera 2 cells. It is readily seen that PDGF-AA does not effect receptor metabolism whereas PDGF-BB induces receptor degradation, thus showing that the PDGF receptors expressed by Tera 2 cells are of the B type. Using the same PDGF receptor antibody as was used in pulse-chase experiments, PDGF receptors were detected in two more teratocarcinoma stem cell lines, i.e. PA-1 (Zuethen et al. 1980) and NTera 2 clD1 (Andrews et al. 1984) (data not shown).

## Discussion

The studies described above show that the expression of growth factors and their receptors is highly regulated during the differentiation of both murine and human teratocarcinoma cell lines. Many more murine cell lines have been investigated than human cell lines and, although there are differences between these lines in the expression of particular growth factors or receptors, in general these cells only express receptors after differentiation but may express and secrete the respective proteins either before or after. A model for growth factor independence of murine EC cell lines based on autocrine growth stimulation may therefore be applicable, with differences between the individual lines being attributable to their different embryonic origin. A similar comparison between different human cell lines is not as yet available. Most studies have been carried out on Tera 2 cl 13 cells or NTera 2 cells, independently cloned derivatives of the Tera 2 cell lines. Although it is

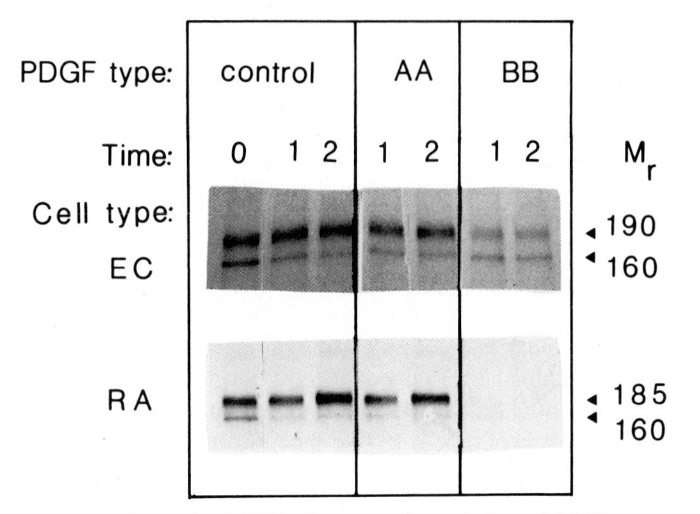


**Fig. 10.** Effect of PDGF isoforms on degradation of PDGF receptors. Monolayers of undifferentiated (*EC*) and differentiated (*RA*) Tera 2 cells were pretreated as described in the legend to Fig. 9. After a 30-min incubation in a methionine- and growth factor-free medium they were labeled for 15 min with [$^{35}$S]methionine. Subsequently they were incubated for 2 h in growth factor-free medium containing methionine (time 0) followed by the indicated chase periods in either the absence or presence of 50 ng/ml PDGF-AA or PDGF-BB. Radiolabeled receptors were precipitated from 1% Triton X-100 extracts using a polyclonal PDGF receptor antibody (Weima et al. 1990). The precipitates were run on a 5%–15% SDS-polyacrylamide gel and prepared for autoradiography. Molecular masses of the proteins indicated by *arrows* are given in kilodaltons

therefore unclear how representative these cells are for human teratocarcinoma cells in general, a striking difference with murine EC cells is their expression of EGF, TGF-β, and PDGF type B receptors prior to differentiation. The presence or absence of PDGF type A receptors remains to be established. Whether this has a functional significance also remains to be established. Cell lines, such as those described in the present studies, provide useful models for identifying the role of growth factors and their receptors in the loss of transformed phenotype during the differentiation of EC cells and may provide insights into the growth and progression of these tumors in vivo.

*Acknowledgments.* We thank Marga van Rooijen, Alie Feyen, and Joop van Zoelen for their enthusiastic participation in these studies.

## *References*

Adamson E, Gaunt SJ, Graham CF (1979) The differentiation of teratocarcinoma stem cells is marked by the types of collagen which are synthesized. Cell 17:469–476

Andrews PW (1984a) Pluripotent embryonal carcinoma clones derived from the human teratocarcinoma cell line Tera-2: differentiation in vivo and in vitro. Lab Invest 50:147–167

Andrews PW (1984b) Retinoic acid induces neuronal differentiation of a cloned human embryonal carcinoma cell line in vitro. Dev Biol 103:285–293

Andrews PW, Bronson DL, Benham F, Strickland S, Knowles BB (1980) A comparative study of eight cell lines derived from human testicular teratocarcinoma. Int J Cancer 26:269–280

Andrews PW, Goodfellow PN, Shevinsky LH, Bronson DL, Knowles BB (1982) Cell surface antigens of a clonal human embryonal carcinoma cell line: morphological and antigenic differentiation in culture. Int J Cancer 29:523–531

Andrews PW, Goodfellow PN, Damjanov I (1983a) Human teratocarcinoma cells in culture. Cancer Surv 2:41–73

Andrews PW, Goodfellow PN, Bronson DL (1983b) Cell surface characteristics and other markers of differentiation of human teratocarcinoma cells in culture. In: Silver LM, Martin GR, Strickland S (eds) Teratocarcinoma stem cells Cold Spring Harbor Laboratory, Cold Spring Harbor, pp 579–590

Bernstine EG, Hooper ML, Grandchamp S, Ephrussi B (1973) Alkaline phosphatase activity in mouse teratoma. Proc Natl Acad Sci USA 70:3899–3902

Biddle C, Li CH, Schofield PN, Tate VE, Hopkins B, Engström W, Huskisson N, Graham CF (1988) Insulin like growth factors (IGFs), and the multiplication of Tera 2, a human teratoma derived cell line. J Cell Sci 90:475–484

Boyd FT, Massague J (1989) Transforming growth factor-β inhibition of epithelial cell proliferation linked to the expression of a 53 kDa membrane receptor. J Biol Chem 264: 2272–2278

Carlin CR, Andrews PW (1985) Human embryonal carcinoma cells express low levels of functional receptors for epidermal growth factor: Exp Cell Res 159:7–26

Cheifetz S, Weatherbee JA, Tsang MLS, Anderson JK, Mole JE, Lucas R, Massague J (1987) The transforming growth factor-β system, a complex pattern of cross-reactive ligands and receptors. Cell 48:409–415

Claesson-Welsh L, Rönnstrand L, Heldin CH (1987) Biosynthesis and intracellular transport of the receptor for platelet-derived growth factor. Proc Natl Acad Sci USA 84:8796–8800

Claesson-Welsh L, Hammacher A, Westermark B, Heldin CH, Nister R (1989) Identification and structural analysis of the A-type receptor for PDGF: similarities with the B-type receptor. J Biol Chem 264:1742–1747

Edwards MKS, Harris JF, McBurney MW (1983) Induced muscle differentiation in an embryonal carcinoma cell line. Mol Cell Biol 3:2280–2286

Engström W, Rees AR, Heath JK (1985) Proliferation of a human embryonal carcinoma-derived cell line in serum-free medium: interrelationship between growth factor requirements and membrane receptor expression. J Cell Sci 73: 361–373

Evans MJ, Kaufman MH (1981) Establishment in culture of pluripotential cells from mouse embryos. Nature 292:154–156

Ewton PZ, Spizz G, Olson EN, Florini JR (1988) Decrease in transforming growth factor-β binding and action during differentiation in muscle cells. J Biol Chem 263:4029–4032

Fogh J, Trempe G (1975) New human tumor cell lines. In: Fogh J (ed) human tumor cell in vitro. Plenum, New York, pp 115–159

Graham CF (1977) Teratocarcinoma cells and normal mouse embryogenesis. In: Sherman MI, Graham CF (eds) Concepts in mammlian embryogenesis. MIT Press, Cambridge, pp 315–394 Mass.

Gronwald RGK, Grant FJ, Haldeman BA, Hart CE, O'Hara PJ, Hagen FS, Ross R, Bowen-Pope DF (1988) Cloning and expression of a cDNA coding for the human platelet-derived growth factor receptor: evidence for more than one receptor class. Proc Natl Acad Sci USA 85:3435–3439

Hart CE, Forstrom JW, Kelly JD, Seifert RE, Smith RA, Ross R, Murray MJ, Bowen-Pope DF (1988) Two classes of PDGF receptor recognize different isoforms of PDGF. Science 240:1529–1531

Heath JK, Deller MJ (1983) Serum-free culture of PC13 murine embryonal carcinoma cells. J Cell Physiol 115:225–230

Heath JK, Shi WK (1986) Developmentally regulated expression of insulin-like growth factors by differentiated murine teratocarcinomas and extraembryonic mesoderm. J Embryol Exp Morphol 95:193–212

Heath JK, Bell SM, Rees AR (1981) The appearance of functional insulin receptors during the differentiation of embryonal carcinoma cells. J Cell Biol 91:293–297

Heldin CH, Bäckström G, Östmen A, Hammacher A, Rönnstrand L, Rubin K, Nister M, Westermark B (1988) Binding of different dimeric forms of PDGF to human fibroblasts: evidence for two separate receptor types. EMBO J 7:1387–1393

Heldin NE, Gustavsson B, Claesson-Welsh L, Hammacher A, Mark J, Heldin CH, Westermark B (1988) Aberrant expression of receptors for platelet-derived growth factor in an anaplastic thyroid carcinoma cell line. Proc Natl Acad Sci USA 85:9302–9306

Hogan B, Taylor A, Adamson E (1981) Cell interactions modulate embryonal carcinoma cell differentiation into parietal or visceral endoderm. Nature 291: 235–237

Jacobowits A, (1986) The expression of growth factors and growth factor receptors during mouse embryogenesis. In: Kahn P, Graf T (eds) Oncogenes and growth control. Springer, Berlin Heidelerg New York, pp 9–17

Jones-Villeneuve EM, McBurney MW, Rogers KA, Kalnins VI (1982) Retinoic acid induces embryonal carcinoma cells to differentiate into neurons and gial cells. J Cell Biol 94:253–262

Martin G (1980) Teratocarcinomas and mammalian embryogenesis. Science 209: 768–775

Martin G (1981) Isolation of a pluripotent cell line from early mouse embryos cultured in medium conditioned by teratocarcinoma stem cells. Proc Natl Acad Sci USA 78:7634–7638

Matsui T, Heidaran M, Niki T, Popescu N, Larochelle W, Kraus M, Pierce J, Aaronson S (1989) Isolation of a novel receptor cDNA establishes the existence of two PDGF receptor genes. Science 243:800–804

Mummery CL, van den Eijnden-van Raaij AJM (1990) Growth factors and their receptors in early murine development. Cell Differ 30:1–18

Mummery CL, van den Brink CE, van der Saag PT, de Laat SW (1984) The cell cycle, cell death and cell morphology during retinoic acid induced differentiation of embryonal carcinoma cells. Dev Biol 104:297–307

Mummery CL, Feijen A, van der Saag PT, van den Brink CE and de Laat SW (1985) Clonal variants of differentiated P19 EC cells exhibit epidermal growth factor receptor kinase activity. Dev Biol 109:402–410

Mummery CL, Feijen A, Moolenaar MH, van den Brink CE and de Laat SW (1986) Establishment of a differentiated mesodermal line from P19 EC cells expressing functional PDGF and EGF receptors. Exp Cell Res 165:229–242

Mummery CL, van Rooijen MA, van den Brink CE, de Laat SW (1987) Cell cycle analysis during retinoic acid induced differentiation of a human embryonal carcinoma-derived cell line. Cell Differ 20:153–160

Mummery CL, van den Eijnden-van Raaij J, Feijen A, Tsung H-C, Kruijer W (1989) Regulation of growth factors and their receptors in early murine embryogenesis. In: deLaat SW, Bluemink JG, Mummery CL (eds) Cell to cell signals in mammalian development. Springer, Berlin Heidelberg New York, pp 231–245 (NATO ASI Series H, vol 26)

Mummery CL, Slager H, Kruijer W, Feijen A, Freund E, Koornneef I, van den Eijnden-van Raaij AJM (1990) Expression of transforming growth factor $\beta_2$ during differentiation of murine embryonal carcinoma and embryonic stem cells. Dev Biol 137:161–170

Rayner M, Graham CF (1982) Clonal analysis of the change in growth phenotype during embryonal carcinoma cell differentiation. J Cell Sci 58:331–344

Rees AR, Adamson ED, Graham CF (1979) Epidermal growth factor receptors increase during the differentiation of embryonal carcinoma cells. Nature 281: 309–311

Rizzino A (1987) Appearance of high affinity receptors for type-β transforming growth factor during differentiation of murine embryonal carcinoma cells. Cancer Res 47:4386–4390

Rizzino A, Bowen-Pope DF (1985) Production of PDGF-like growth factors by embryonal carcinoma cells and binding of PDGF to their endoderm-like differentiated cells. Dev Biol 110:15–22

Rizzino A, Crowley C (1980) Growth and differentiation of embryonal carcinoma cell line F9 in defined media. Proc Natl Acad Sci USA 77:457–461

Scatchard G (1949) The attraction of proteins for small molecules and ions. Ann NY Acad Sci 51:660–672

Segarini PR, Seyedin SM (1988) The high molecular weight receptor to transforming growth factor-β contains glycosaminoglycan chains. J Biol Chem 263:8366–8370

Shevinsky LH, Knowles BB, Damjanov I, Solter D (1982) Monoclonal antibody to murine embryos defines a stage-specific embryonic antigen expressed on mouse embryos and human teratocarcinoma cells. Cell 30:697–705

Solter D, Knowles BB (1978) Monoclonal antibody defining a stage-specific mouse embryonic antigen (SSEA-1). Proc Natl Acad Sci USA 75:5565–5569

Strickland S, Mahdavi V (1978) The induction of differentiation in teratocarcinoma stem cells by retinoic acid. Cell 15:393–403

Thompson S, Stern PL, Webb M, Walsh FS, Engström W, Evans EP, Shi W-K, Hopkins B, Graham CF (1984) Cloned human teratoma cells differentiate into neuron-like cells and other cell types in retinoic acid. J Cell Sci 72:37–64

Van Zoelen EJJ (1989) Receptor-ligand interaction: a new method for determining binding parameters without a priori assumptions on non-specific binding. Biochem J 262:549–556

Van Zoelen EJJ, Koornneef I, Holthuis JCM, Ward-van Oostwaard TMJ, Feijen A, de Poorter TL, Mummery CL, van Buul-Offers CS (1989) Production of insulin-like growth factors, platelet derived growth factor, and transforming growth factors and their role in the density-dependent growth regulation of a differentiated embryonal carcinoma line. Endocrinology 124:2029–2041

Wang S-Y, Gudas LJ (1983) Isolation of cDNA clones specific for collagen IV and laminin from teratocarcinoma cells. Proc Natl Acad Sci USA 80:5880

Weima SM, van Rooijen M, Mummery CL, Feijen A, Kruijer W, de Laat SW, van Zoelen EJJ (1989a) Differentially regulated production of platelet-derived growth factor and of transforming growth factor beta by a human teratocarcinoma cell line. Differentiation 38:203–210

Weima SM, van Rooijen MA, Feijen A, Mummery CL, van Zoelen EJJ, de laat SW, van den Eijnden-van Raaij AJM (1989b) Transforming growth factor-β and its receptor are differentially regulated in human embryonal carcinoma cells. Differentiation 41:245–253

Weima SM, Stet LH, van Rooijen MA, van Buul-Offers SC, van Zoelen EJJ, de Laat SW, Mummery CL (1989c) Human teratocarcinoma cells express functional insulin-like growth factor I receptors. Exp Cell Res 184:427–439

Weima SM, van Rooijen MA, Mummery CL, Feijen A, de Laat SW, van Zoelen EJJ (1990) Identification of the type-B receptor for platelet-derived growth factor in human embryonal carcinoma cells. Exp Cell Res 186:324–331

Williams LT (1989) Signal transduction by the platelet-derived growth factor receptor. Science 243:1564–1570

Yarden Y, Escobedo JA, Kuang WJ, Yang-Feng TL, Daniel TO, Tremble PM, Chen EY, Ando ME, Harkins RN, Francke U, Fried VA, Ulrich A, Williams LT (1986) Structure of the receptor for platelet-derived growth factor helps define a family of closely related growth factor receptor. Nature 323:226–232

Zuethen J, Nørgaard JOR, Avner P, Fellow M, Wartiovaara J, Vaheri A, Rosen A, Giovanella BC (1980) Characterization of a human ovarian teratocarcinoma-derived cell line. Int J Cancer 25:19–32

# *Growth Factors in Human Germ Cell Cancer*

W.H. Miller, Jr. and E. Dmitrovsky

Laboratory of Molecular Medicine, Department of Medicine,
Memorial Sloan-Kettering Cancer Center, Cornell University Medical College,
New York, NY 10021, USA

## Introduction

A growing body of evidence suggests that malignancy often represents a failure of differentiation. Several model systems have been developed that show an inverse relationship between differentiation and malignancy. As tumor cells differentiate, they typically lose proliferative potential and the ability to cause tumors when implanted into susceptible hosts. Thus, an understanding of the cellular mechanisms of differentiation may yield insight into the mechanisms of the abnormal regulation of growth by cancer cells. Murine and human models of differentiation of tumor cells, including mouse erythroleukemia (MEL) cells, human leukemic cell lines, and mouse F9 teratocarcinoma cells, have been used to explore changes in the expression of important growth-regulated genes, such as cellular oncogenes and growth factors, as a function of differentiation induced in vitro. For example, a biphasic decline of c-*myc* and c-*myb* mRNA levels has been seen following dimethyl sulfoxide (DMSO) induction of MEL cells (Ramsey et al. 1986). Indeed, some authors have suggested a causal relationship between the downregulation of oncogene products and both differentiation and loss of malignancy. This has been supported by experiments showing that expression of transfected c-*myc* or c-*myb* genes into MEL cells blocks the induction of differentiation by DMSO (Dmitrovsky et al. 1986; Clarke et al. 1988). Differentiation of F9 cells can be induced by retinoic acid (RA) and dibutyryl cAMP and, as in MEL cells, is accompanied by specific changes in expression of protooncogenes, biochemical markers, developmentally active genes, and growth factors (reviewed in Miller et al. 1990a). These changes correlate with a decline in the malignant and proliferative potential of these differentiated cells (Strickland and Sawey 1980; Strickland and Mahdavi 1978).

Recent Results in Cancer Research, Vol. 123

**Retinoic Acid Induced Differentiation**

The human teratocarcinoma line NTERA-2D1 (NT2/D1) provides a useful model to extend these studies of oncogene regulation during RA-induced differentiation to a human germ cell tumor. NT2/D1 cells have the capability to differentiate into a variety of tissue types in vivo when injected into athymic nude mice and can be induced to differentiate along a neuronal pathway in vitro after exposure to RA (see Andrews 1988). Differentiation induced by RA leads to reduced stage-specific embryonic antigen-3 (SSEA-3) expression in differentiated cells along with an increase in the expression of the preneuronal marker A2B5 (Andrews 1988).

Recent work in our laboratory has correlated induced differentiation with specific changes in mRNA expression of a panel of over 40 cellular oncogenes, growth factors, and markers of germ cell tumor differentiation (Miller et al. 1990a,b). The steady-state levels of total cellular RNA in NT2/D1 cells after 6 days of RA treatment were compared with levels of RNA in uninduced NT2/D1 cultures. Many genes showed no detectable level of expression or a level of expression that is unchanged by exposure to RA. Table 1 reviews the results for a few such unregulated genes and the important regulated genes. Tissue plasminogen activator (TPA) and laminin, upregulated in F9 cells (Strickland and Mahdavi 1978; Wang and Gudas 1983), were expressed but unregulated in NT2/D1 cells. As reported by others (Hauser et al. 1985; Krust et al. 1989), expression of several homeobox-containing genes and the nuclear RA receptor RAR-β was induced by RA in these human teratocarcinoma cells. Several nuclear protooncogenes were modestly downregulated during differentiation. While c-*myc* and N-*myc* are markedly downregulated in murine teratocarcinoma cells induced to differentiate (Finkelstein and Weinberg 1988), their expression in NT2/D1 was only variably and modestly decreased. This is consistent with the report (Schofield et al. 1987) that c-*myc* expression is unchanged by RA-induced differentiation of the human teratocarcinoma cell line Tera 2 cl 13. In contrast, two growth factors, *TGF*-α and *hst-1*/*kFGF*, were prominently downregulated by RA treatment. Both genes were unexpressed at the level of total cellular RNA Northern analysis within 48–96 h of RA treatment. This loss of *TGF*-α and *hst-1*/*kFGF* expression correlated with a loss of proliferative potential and tumorigenicity of RA-treated NT2/D1 cells (Dmitrovsky et al. 1990; Miller et al. 1990a). Induction of differentiation by RA resulted in decreased cloning efficiency by limiting dilution and in soft agar (anchorage-independent growth). Injection of RA-treated NT2/D1 cells into nude mice was also associated with a marked decline in tumor formation compared with injection of untreated NT2/D1 cells.

The finding of two prominently downregulated genes, *TGF*-α and *hst-1*/*kFGF*, suggests they may play a role in the growth and differentiation of the NT2/D1 cell. To explore this possibility, we initially focused our attention on the *TGF*-α protein product. We added purified *TGF*-α protein to soft agar

**Table 1.** Effect of 6 days RA treatment on total cellular RNA expression of growth factors, protooncogenes, and markers of differentiation in NT2/D1 cells, compared with uninduced cultures

| Probe | Unexpressed | No Change | Decreased | Increased |
|---|---|---|---|---|
| c-*myc* | | | +/− | |
| TPA | | + | | |
| GAPDH | | + | | |
| β-actin | | + | | |
| Laminin | | + | | |
| *Hst*-1/kFGF | | | ++ | |
| *TGF*-α | | | ++ | |
| *HOX2.1* | | | | ++ |
| *HOX2.2* | | | | ++ |
| RAR-β | | | | ++ |

+ denotes a decrease in expression that is consistently greater than 50% but less than 90% as determined by densitometric scanning of autoradiographs from Northern filters from separate inductions; ++ denotes complete absence of RNA expression detectable by Northern analysis of total cellular RNA of RA-treated cells (decrease) or an absence of RNA expression in the untreated cells (increase).

colonies of NT2/D1 cells grown under limited fetal calf serum (FCS) conditions and found that *TGF*-α can augment the soft agar growth of these cells (Dmitrovsky et al. 1990). A related growth factor, epidermal growth factor (*EGF*), which shares with *TGF*-α the ability to bind the EGF receptor, can substitute for this growth stimulatory effect. Fig. 1 shows this dose-dependent stimulation of growth. To provide further support for the specific action of *TGF*-α via the EGF receptor, we found that a blocking monoclonal antibody directed against the EGF receptor can inhibit the augmentation of soft agar growth by *TGF*-α or *EGF* (Dmitrovsky et al. 1990). However, there is no augmentation of soft agar cloning at any tested dose of *TGF*-α when the FCS concentration of the culture medium is below 5% (Dmitrovsky et al. 1990). Similarly, Engstrom reported that *EGF* is not mitogenic in Tera-2-derived cells grown in serum-free medium (Engström et al. 1985). These results imply that some component of FCS is required in addition to *TGF*-α or *EGF* for the sustained growth of these cells in culture. Our RNA data suggest the possibility that one dose-limiting component missing in low-serum or serum-free media might be the downregulated growth factor *hst-1/kFGF*. Experiments to test this hypothesis are under way.

## Hexamethylene Bisacetamide Induced Differentiation

We sought to extend the above observations to explore whether the changes in gene expression that accompany differentiation induced by RA are general features of human embryonal cancer cell differentiation. First, we studied

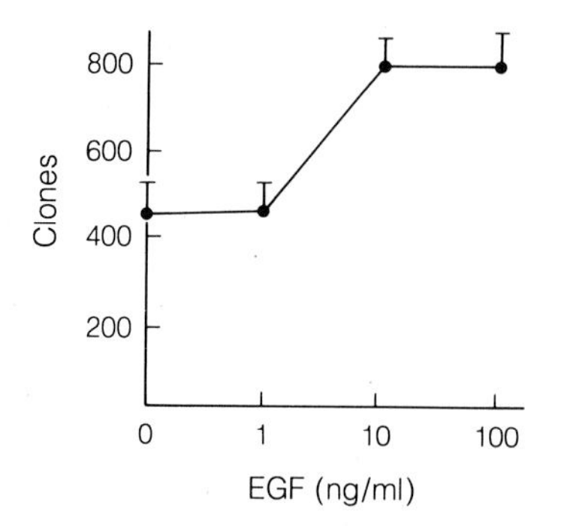


**Fig. 1.** Dose-dependent effect of the addition of EGF protein on the soft agar cloning of NT2/D1 cells grown in DME medium containing 5% FCS. The number of soft agar clones per square centimeter field are shown. Colonies greater than 50 μm in size are scored 3 weeks after plating

changes in gene expression in NT2/D1 cells induced by hexamethylene bisacetamide (HMBA) to differentiate along a different pathway. HMBA has been shown to induce a distinct pathway of differentiation in NT2/D1 cells as evidenced by a different morphology and the induction of specific surface markers (Andrews 1988), while it causes a loss of proliferative capacity and tumorigenicity equal to that induced by RA (Miller et al. 1990b). HMBA does not induce expression of three genes induced by RA, the β-nuclear RA receptor or the homeobox-containing genes, *Hox 2.1* and *Hox 2.2* (Miller et al. 1990b). This provides further evidence that the RA and HMBA phenotypes of these cells are distinct and suggests different second messenger pathways.

Most interestingly, in spite of these differences, HMBA also causes downregulation of *TGF-α* and *hst-1/kFGF*. We found that both RA- and HMBA-treated cells have a marked decline in cloning efficiency (compared with untreated NT2/D1 cells by soft agar and limiting dilution clonal assays) and tumorigenicity (after injection into nude mice) (Miller et al. 1990b). Thus, different chemical inducers, which stimulate distinct differentiation pathways, cause a common growth factor downregulation in the NT2/D1 cell and lead to a similar loss of malignant potential.

## Model of Chemically Induced Differentiation of EC Cells

We propose a model to portray the pathways of differentiation in NT2/D1 cells, shown in Fig. 2. We and others have shown common intermediate events to both pathways: downregulation of growth factors and of embryonal carcinoma surface markers. We find both common and distinct late events: loss of cloning efficiency and tumorigenic potential is common to both inducers while the induced cell surface markers differ (Miller et al. 1990b;

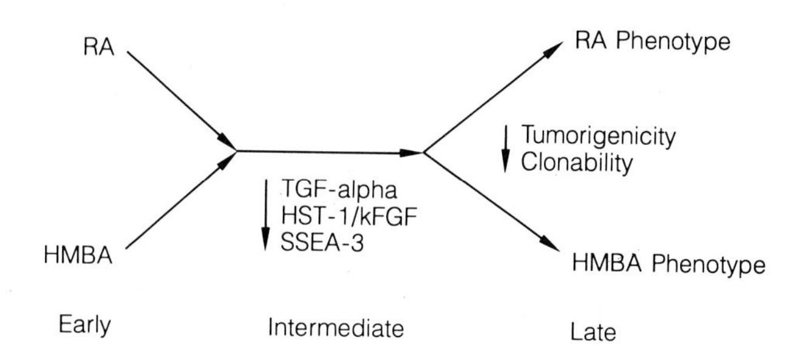


**Fig. 2.** A model is presented to highlight similarities and differences between the two pathways of differentiation of NT2/D1 cells. *Early*, *intermediate*, and *late* refer to events occurring within 24 h, 1–4 days, or after morphologic differentiation becomes apparent. Although little is known about the second messenger pathway and early gene activation of RA and HMBA, common intermediate events are shown, as are both common and distinct late events

Andrews 1988). Less is known about the early events stimulated by either inducer. HMBA has been shown to induce a rapid (30–60 min) translocation of protein kinase C (PKC) activity to the membrane in one cell line (Melloni et al. 1987). Three specific nuclear receptors for RA have recently been cloned. Of these, it was recently reported that RAR-β is prominently induced by RA treatment in F9 teratocarcinoma cells and in human Tera 2 cl 13 teratocarcinoma cells (Hu and Gudas 1990; Krust et al. 1989). We have found that RAR-β is induced by RA treatment but not by HMBA treatment in NT2/D1 cells. These preliminary studies further suggest that RA and HMBA act via different early events. More research on the early actions of these inducers is clearly needed.

The finding that two growth factor genes are prominently downregulated in both pathways of differentiation suggests that they may be involved in the differentiation program and/or the maintenance of the malignant growth of one embryonal carcinoma cell line, NT2/D1. We probed a panel of human teratocarcinoma cell lines for expression of *TGF-α* and *hst-1/kFGF* and found both growth factors to be expressed in the majority. Studies of growth factor expression in primary tumor samples are in progress. To explore further the link between the downregulation of these growth factors and differentiation, we studied growth factor regulation by RA and HMBA in a human embryonal cancer cell line, N2102ep. N2102ep expresses *TGF-α* and *hst-1/kFGF* in the uninduced state and can be induced to morphologic differentiation in response to HMBA but not to RA. We found that growth factor expression was downregulated only in response to the effective inducer of differentiation, HMBA (Miller et al. 1990b).

## Conclusion

From the data presented here, we hypothesize that the expression of these two growth factors might be important in the growth or differentiation of NT2/D1 or other germ cell cancer cell lines and even primary germ cell cancers. These studies may provide a scientific rationale for a treatment strategy in refractory human germ cell cancer based on growth factor antagonists. These studies may also lead to the correlation of growth factor expression with the clinical course of germ cell cancer. We hope that these efforts will ultimately result in the identification of unique biological features of germ cell cancer which can be exploited for the benefit of these cancer patients.

*Acknowledgements.* This work was supported in part by a Lederle Fellowship to Dr. Miller, the American Cancer Society Clinical Oncology Career Development Award #89–129 to Dr. Dmitrovsky, and the American Cancer Society Grant PDT #381.

## *References*

Andrews PW (1988) Human teratocarcinomas. Biochim Biophys Acta 948:17–36

Clarke MF, Kukowska-Latallo JF, Westin E, Smith M, Prochownik E (1988) Constitutive expression of a c-*myb* cDNA blocks friend murine erythroleukemia cell differentiation. Mol Cell Biol 8:884–892

Dmitrovsky E, Kuehl WM, Hollis GF, Kirsch IR, Bender TP, Segal S (1986) Expression of a transfected human c-*myc* oncogene inhibits differentiation of a mouse erythroleukemia cell line. Nature 322:748–750

Dmitrovsky E, Moy D, Miller WH Jr, Li A, Masui H (1990) Retinoic acid causes a decline in TGF-α expression, cloning efficiency, and tumorigenicity in a human embryonal cancer cell line. Oncogene Res 5:233–239

Engström W, Rees AR, Heath JK (1985) Proliferation of a human embryonal carcinoma-derived cell line in serum-free medium: interrelationship between growth factor requirements and membrane receptor expression. J Cell Sci 73: 361–373

Finkelstein R, Weinberg R (1988) Differential regulation of N-*myc* expression in F9 teratocarcinoma cells. Oncogene Res 3:287–292

Hauser CA, Joyner AL, Klein RD, Learned TK, Martin GR, Tjian R (1985) Expression of homologous homeo-box-containing genes in differentiated human teratocarcinoma cells and mouse embryos. Cell 43:19–28

Hu L, Gudas LJ (1990) Cyclic AMP analogs and retinoic acid influence the expression of retinoic acid receptor α, β, and gamma mRNAs in F9 teratocarcinoma cells. Mol Cell Biol 10:391–396

Krust A, Kastner P, Petkovich M, Zelent A, Chambon P (1989) A third human retinoic acid receptor, hRAR-gamma. Proc Natl Acad Sci USA 86:5310–5314

Melloni E, Pontremoli S, Michetti M, Sacco O, Cakiroglu AG, Jackson JF, Rifkind RA, Marks PA (1987) Protein kinase C activity and hexamethylene bisacetamide induced erythroleukemia cell differentiation. Proc Natl Acad Sci USA 84: 5282–5286

Miller WH Jr, Moy D, Li A, Grippo JF, Dmitrovsky E (1990a) Retinoic acid induces down-regulation of several growth factors and proto-oncogenes in a human embryonal cancer cell line. Oncogene 5:511–517

Miller WH Jr, Kurie J, Moy D, Lucas DA, Grippo JF, Masui H, Dmitrovsky E (1990b) Growth factor downregulation is linked to differentiation and loss of tumorigenicity in human teratocarcinoma cells. To be published

Ramsay RG, Ikeda K, Rifkind RA, Marks PA (1986) Changes in gene expression of erythroleukemia protooncogenes, globin genes, and cell division. Proc Natl Acad Sci USA 83:6849–6853

Schofield PN, Engstrom W, Lee AJ, Biddle C, Graham CF (1987) Expression of c-*myc* during differentiation of the human teratocarcinoma line Tera-2. J Cell Sci 88:57–64

Strickland S, Mahdavi V (1978) The induction of differentiation in teratocarcinoma stem cells by retinoic acid. Cell 15:393–403

Strickland S, Sawey MJ (1980) Studies on the effect of retinoids on the differentiation of teratocarcinoma stem cells in vitro and in vivo. Dev Biol 78:76–85

Wang S-Y, Gudas LJ (1983) Isolation of cDNA clones specific for collagen IV and laminin from mouse teratocarcinoma cells. Proc Natl Acad Sci USA 80:5880–5884

# *Subject Index*

Printing and binding: Druckerei Triltsch, Würzburg